Hierarchy Theory

Hierarchy Theory

A Vision, Vocabulary, and Epistemology

Valerie Ahl and T. F. H. Allen

Illustrated by Paula Lerner

Columbia University Press

New York

Columbia University Press
New York   Chichester, West Sussex

Copyright © 1996 Columbia University Press

Library of Congress Cataloging-in-Publication Data

Ahl, Valerie.
Hierarchy theory : a vision, vocabulary, and epistemology /
Valerie Ahl and T. F. H. Allen.
p.   cm.
Includes bibliographical references and index.
ISBN 0-231-08480-3 (cloth). — ISBN 0-231-08481-1 (pbk.)
1. Science—Philosophy.   2. Science—Methodology.   3. Problem
solving.   I. Allen, T.F.H.   II. Title.
Q175.A334   1996
003'.7—dc20   96-174
                                                          CIP

∞

Casebound editions of Columbia University Press books
are printed on permanent and durable acid-free paper.

Printed in the United States of America
c 10 9 8 7 6 5 4 3 2 1

O chestnut-tree, great-rooted blossomer,
Are you the leaf, the blossom or the bole?
O body swayed to music, O brightening glance,
How can we know the dancer from the dance?

W. B. YEATS, "AMONG SCHOOLCHILDREN"

For Peter, Joan, Crokie, Frank, and little Josephine

—We miss you, Peter

Contents

Acknowledgments

While Paula Lerner is formally acknowledged as a collaborator on this project, her generosity in getting involved and producing her wonderful photographs deserves more than the mere citations that appear in the figure legends. She is a professional free-lance photographer who owns the copyright to all her images here and the many others that we almost used but that were squeezed out in some hard choices. Readers wishing to contact her can use the address 30 Selwyn Road, Belmont, Massachusetts 02178; telephone (617) 489-6747.

There have been many friends over the years who have read early drafts of this book and offered critiques. Our thanks to them all. T. F. H. Allen's Introduction to Systems Biology classes used those early drafts, and our thanks go out to the students there who offered us advice. Colleagues and friends who read the manuscript and helped us a lot with their extended criticism include Jonas Langer, Elliot Turiel, Doc Uckuck, Louise Comas, Keith Anderson, and Virginia Slaughter. There were many others too numerous to name who gave us insights and comments, and we thank them all. Mark Groves of Harvey Mudd College gave us the Yeats quotation that opens this book. Gene Robkin was our source on the anecdote about the American fighter squadron.

Kandis Elliot appears on the legends of figures that she crafted, but those citations do not do justice to the enormous effort that she put into that work. We are grateful to the Botany Department of the University of Wisconsin for its support of Kandis's graphics laboratory and the photo lab. Claudia Lipke performed the conversion of slides and other materials to prints for many of the figures.

The research behind this book was supported in part by NSF award DEB 9215020 to T. F. H. Allen at the University of Wisconsin, and NSF award BSR 90-11660 through the Center for Limnology at the University of Wisconsin, and a summer fellowship to Valerie Ahl from the Institute of Human Development at the University of California at Berkeley.

Hierarchy Theory

1

Confronting the Complexity
of Our Time

The Problem

In all ages humanity has been confronted by complex problems. The difference between then and now is that contemporary society has ambitions of solving complex problems through technical understanding. Of course, the heroes of the industrial age, with their bridges and engines, were spectacular. But contemporary achievements, from actual moon shots to creating Hollywood space sagas, involve creativity in the face of a new order of complexity. The very character of the modern world is that we are prepared to confront issues of complexity head-on.

While science and society have not exactly tamed complex systems, there has been undeniable progress. Two main strategies are responsible for these advances; one action-oriented and the other more thoughtful. The first strategy is to reduce complex problems by gaining tight control over behavior. It is a mechanical solution in the style of differential equations and Newtonian calculus. The second solution is to expand the problem domain to include the observer as well as the observed. It is focused on context and is in the spirit of Jaynes's calculus (as discussed by Rosenkrantz). The approach here is not a frontal assault on the problem directly, using mechanical solutions. Rather it is an analysis of the process of problem-solving so that it can be tailored to the specific situation at hand.

Ineffective attempts at solving complex problems often turn on repeating a small number of mistakes invited by taking the process of observation for granted. The approach we take challenges such things as why a problem is considered interesting, how factors are delimited, and

THE SUSPENSION BRIDGE, NEAR BANGOR.

CARNARVONSHIRE.

Drawn & Engraved for DUGDALES ENGLAND & WALES Delineated

FIGURE 1.1   "Of course, the heroes of the industrial age, with their bridges and engines, were spectacular."

what sorts of solutions would be accepted. Such challenges are the operationalization of the second, more thoughtful approach that expands the problem domain.

Centuries ago, the only sensible thing to do was to capitulate to enormously complex situations and to deal with them through ritual. Jacob Bronowski in "The Ascent of Man" points to the ritual of the Samurai swordsmith as one device for achieving the precision required to make a spectacularly sharp edge using only blunt tools. The master exquisitely heats, folds, beats, and quenches the blade repeatedly. Thus the ordering of millions of microscopic crystals at the edge of the blade is controlled with an instrument as coarse as a hammer.

Other rituals may be purely religious and involve surrender to some belief system. For example, the Romans had gods for various agricultural practices, such as Repactor, the god of second plowing, and Sterculius, the god of manuring. If Roman ritual plowing did not have the desired

FIGURE 1.2 "Contemporary achievements, from actual moon shots to creating Hollywood space sagas, involve creativity in the face of a new order of complexity." *Drawn by Kandis Elliot.*

effect on the crops, then the farmers presumed that the rite had not been performed with enough care, and they did it again with greater attention to detail. Ritual codifies all the steps of the dance, thus preserving the ones that matter. Rituals are effective even if the guardians of knowledge have no idea which detailed actions actually do make a difference. In the absence of understanding, a good strategy is: do what worked last time.

Rituals allow society to deal with complexity beyond what can be explicitly comprehended. Ritual is one example of a mechanical approach to complex problems. In the mechanical approach, the problem space is reduced to a small set of factors, over which tight control is exercised. The mechanical approach has led to many modern successes. Contemporary high revolution automobile engines are possible because

FIGURE 1.3   "The ritual of the Samurai swordsmith as one device for achieving the precision required to make a spectacularly sharp edge using only blunt tools." *Photograph by Paula Lerner.*

FIGURE 1.4 "The engines of prior eras, such as a Stevenson steam engine, or even a '54 convertible, operate with many more degrees of freedom."
Drawn by Kandis Elliot.

of precise relations among parts. The engines of prior eras, such as a Stevenson steam engine, or even a '54 convertible, operate with many more degrees of freedom. Today's machines achieve precision through much narrower tolerances than the machines of earlier times.

Many situations can be effectively described with a chaotic model. In chaos, minute differences in initial conditions lead to radically different outcomes quite quickly. There are two ways to fight this tendency of each instance giving a different result. Both are effective but neither is very aesthetically satisfying. The first way is to avoid letting things run for very long without stepping in to make some correction. Flights to the moon are achieved in this fashion: the original thrust away from the Earth's gravitational sphere is never intended as a single shot, because engineers know that no matter how hard they try, a course correction will be needed midjourney. This solution depends on reducing one large prob-

lem to two smaller problems: escaping Earth's gravity, and aiming precisely at the moon. A second way to fight chaos is to try very hard to get the starting configuration, technically called initial conditions, exactly the same every time the process is started. This solution never works in the long run, but it can postpone unpredictable behavior for a while. Champion billiards players apply this principle. By striking the cue ball with exquisite precision they get the balls to go in the right pockets. Demanding narrow tolerances is an attempt to avoid large-scale effects of minute differences at each particular outset.

The moderate success that has been achieved by narrowing tolerances implies an understanding that is often illusory. For all our modern computing power and technical wizardry, the standard strategy is still some version of "doing what worked last time" and codifying it in ritual. For example, the United States was saved from the ravages of Thalidomide not by drug trial research on limb bud impairment but by a ritual process of drug approval by the Food and Drug Administration. It was known that the drug could cause tingling in the fingers and toes, a telltale sign that was ignored in Europe. The manufacturers knew of some effects on embryonic limb growth in test animals, but results were reported to emphasize species where there were no such problems. In the face of the ritual process of drug approval in the United States, vested interests that were happy to take shortcuts in Europe could not get their way in America. Ritual is very important in stopping quick and dirty solutions. What makes ritual so effective is that it does not succumb to rational argument, erected in favor of political or economic expedients. Religious ritual blunts rational objections in exactly this way.

Instead of changing actual understanding, modern technology's contribution is more often to change how we approach complexity. It is arguably a change in attitude rather than in competence. Take, for example, acid rain. Acid rain is not a particularly modern problem. The unspoiled Lake District of England has in fact been devastated by northern industrial pollution for two hundred years, such that many small lakes have no fish. Until very recently the acidity was put down to natural processes in bogs. Modern measuring devices to detect patterns of acidity are helpful, but they have merely drawn our attention to the fact that

---

FIGURE 1.5   "The unspoiled Lake District of England has in fact been devastated by northern industrial pollution for two hundred years."
Print of W. Heaton Cooper, a celebrated Lakeland watercolorist.

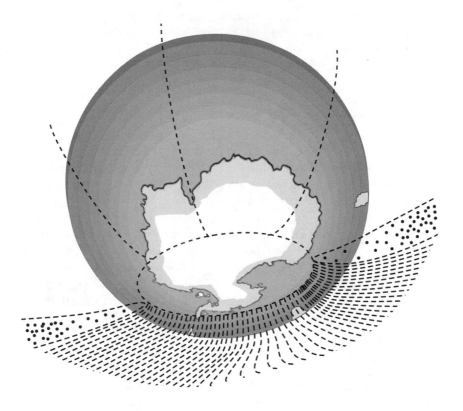

FIGURE 1.6 The hole in the ozone layer over Antarctica. Chloroflourocarbons from refrigerators have transported chlorine to the stratosphere, where it catalytically destroys ozone much faster than by natural processes. A stable convection cell in the darkness of the Antarctic winter stops ozone from the tropics from replacing the lost ozone. *Drawn by Kandis Elliot*

acid rain is an old, not a new, problem. Knowing the source of acid rain is only a small step toward a solution. What to do about it is still an open question, with larger sociological and political elements rather than strictly technical answers. It is more an issue of being prepared to pay the price to remedy acid rain than one of ignorance of cause and effect.

Sometimes a genuinely new problem may take modern technology to detect it. In the case of the hole in the ozone layer, computer-based records of skin cancer epidemiology gave the early warning, and remote

FIGURE 1.7 "Predictability requires addressing multiple levels of analysis simultaneously." Stairway in the Vatican. *Photograph by Paula Lerner*

sensing from space graphically showed the source of the problem. Had a hole in the ozone layer appeared in the last century, people would have died from skin cancer, and ultraviolet radiation would have wreaked havoc in the biosphere, without anyone noticing the pattern or determining a cause. The contemporary world would merely have accepted the higher incidences as a natural background parameter. Once again, the effect of technology is to draw attention to problems rather than actually solve them.

The mechanical approach to complex systems has been particularly successful in the domains of computer and mechanical engineering. Modern computers allow very fast processing of what only ten years ago would have been an inordinate amount of data. Analyzing and storing large quantities of data is now trivial. The technological fix should, however, be viewed with caution. The challenge to modern science is not how to deal with data but how to interpret and integrate them. We love our computer hardware, and, as with all love affairs, the one who is enamored has a very biased view. It is fair to say that the human on the other

FIGURE 1.8 "Open the door to unexpected levels of analysis."
Photograph by Paula Lerner

side of the keyboard has received far less attention. Hierarchy theory redresses this imbalance by analyzing the process of observation. Focusing on the observer/observed interface gives insights into a new class of solutions based on pragmatic assessment of the limits of understanding rather than on ritual or tight control within a limited problem domain. Driving toward tight material control involves much focused intellectual creativity, but it takes the goals of the endeavor for granted. The alternative strategy for taming complexity that we employ is to evaluate the interaction between the question and the questioner. It is an intellectual refinement, in which the goals and motives of the investigator are examined before any remedy is sought. Such a systems approach involves including the observer, and human values, in the process of coming to terms with complex problems.

Complex systems are defined as those systems which require fine details to be linked to large outcomes. Forging this link in a way that allows predictability requires addressing multiple levels of analysis simultaneously. Examples are as diverse as a telecommunications satellite or a poem. Hierarchy theory's chief focus is on levels of analysis as they relate to understanding. The theory uses insights about the role of the observer to identify limits to understanding. It addresses the nature of questions and determines that some of them are so inept that they have no answer, no matter how much we might want one. Hierarchy theory uncovers hidden necessities so that, as a society, we can avoid tilting at windmills and turn toward formidable, but beatable, opponents.

The scientist who uses hierarchy theory is often less interested in finding out what is "really" happening in the material system, and focuses more on finding a powerful point of view. A hard-line hierarchist might even go so far as to say that scientific discovery is always of points of view, and only trivially a matter of material verity. Anyone uncomfortable with that position can still use hierarchy theory to effect, while retaining a more realist philosophy. It is probably no accident that hierarchy theory has emerged in the same intellectual climate that spawned deconstructivist perspectives, for both open the door to unexpected levels of analysis.

In science, hierarchy theory is the child of a cross-fertilization of traditional disciplines. One of its founding fathers, Herbert Simon, received a Nobel Prize for economics. Another Nobel Laureate hierarchist, Ilya Progogine, hails from chemistry. He identified what happens when new hierarchical levels emerge, not just for chemical reactions but as a gen-

SUBJECT   WORLD

acts   Boundary   reacts

Interaction = source of Order

Development = progressive change
in subject as function of
interaction with the world

eral process. Ecologists are an equally distinct group that uses hierarchy theory and applies it to practical problems. Yet another related field is psychology. Jean Piaget's Constructivist theories of genetic epistemology are uniquely hierarchical in focus. In fact, late in life, Piaget indeed studied complex hierarchical systems with Prigogine, the chemist. In all these fields, hierarchy theory is less a theory to be tested than a position of self-consciousness of the investigator as to how the pursuit of understanding proceeds. There is a central utilitarian component to the theory which is not found in psychology or philosophy at large. Hierarchy theory has close neighbors, yet it maintains its own particular turf.

Psychology has made enormous strides in giving a new level of understanding to both human perception and learning. Hierarchy theory addresses similar issues, but lies between perception and learning to focus on observation. Observation rests on perception, but it also draws on prior learning, and is a means for gaining new knowledge. Observation requires not only seeing but looking. As Louis Pasteur remarked, "Chance favors the prepared eye." In order to make an observation one must have an idea of what could be seen, and a framework of beliefs into which new observations, both confirming and disconfirming, may be interwoven. Thus observation is the interface between perception and learning. Prior learning is invoked to structure new perception, and new perceptions are used to advance learning in the form of theory construction and modification. The interrelations between observation, perception, and learning is the substance of hierarchy theory, and the focus of this book.

FIGURE 1.9 Piaget argues, and we agree, that knowledge comes not from the external world or from the observer but from interaction between observer and the world. A nine-year-old boy kept from interacting with the world by a bubble to protect him from infection could not develop a normal understanding of the view from his window. He would ask if the buildings he could see were more than surfaces and whether they had windows in back. One day a fog helped him comprehend the explanation that he had been given before but had barely believed and had not properly understood—that the smaller buildings were actually bigger but looked smaller because they were farther away. Knowledge comes from interaction. The account of the boy in the bubble is from Mary Murphy and Jacqueline Vogel in Developmental and Behavioral Pediatrics 6 (1985): 118–21. Looking out from the isolator: David's perception of the world. *Figure drawn by Kandis Elliot*

Clarifying Questions

Science usually settles issues by testing an idea against observations of a material system. This approach to problem solving has a history of working so well that another class of solution is often overlooked. Instead of focusing on how to limit or control the problem space in order to achieve predictability, hierarchy theory is useful in finding alternative ways of framing a question. In many problematic situations, the challenge is to find a new perspective that works rather than to acquire more data within an old framework. In such cases, a solution cannot be found by making more observations, because the contradiction, disagreement, or uncertainty does not reside in some unresolved aspect of the material system. Looking again or more closely will not help. There is not an insufficiency of data but rather some internal conflict in the way one or more observers have chosen to look at the world. In these cases, the heart of the problem resides in an inadequate question.

Take, for example, semantic arguments that stem from the mistaken coupling of symbols to the objects they represent and assuming a material link. The following translation from the Russian psychologist Lev Vygotsky illustrates this error.

> We all know the old story about the rustic who said he wasn't surprised that savants with all their instruments could figure out the size of stars and their course—what baffled him was how they found out their names. Simple experiments show that preschool children "explain" the names of objects by their attributes. According to them, an animal is called "cow" because it has horns, "calf" because its horns are still small, "dog" because it is small and has no horns; an object is called "car" because it is not an animal. When asked whether one could interchange the names of objects, for instance call a cow "ink," and ink "cow," children will answer no, "because ink is used for writing, and the cow gives milk." An exchange of names would mean an exchange of characteristic features, so inseparable is the connection between them in the child's mind. In one experiment, the children were told that in a game a dog would be called "cow." Here is a typical sample of questions and answers:
>
> *"Does a cow have horns?"*
> "Yes."

FIGURE 1.10   "Does a cow have horns?" Photograph by Paula Lerner.

"But don't you remember that the cow is really a dog? Come now, does a dog have horns?"
"Sure, if it is a cow, if it's called cow, it has horns. That kind of dog has got to have little horns." (Thought and Language, p. 129)

Mistaking incidental features, such as names, for inherent properties of objects is a common error found in young children's reasoning. This phenomenon was labeled nominal realism by Piaget. When these errors occur in the naive they seem ridiculous, but sophisticates have their own version of the same problem.

When scholars argue over the "true" definition, rather than the most useful definition, they might as well be trying to find the names of the stars. Semantic arguments are a regular part of science, and they derive from an overcommitment to certain names or definitions. Sometimes the conflict arises because the same name means different things to the respective conversants.

The frequency of nominal realist arguments within a discipline appears to be related to whether the subject matter is tangible or intan-

FIGURE 1.11 "Sure, if it is a cow, if it's called cow, it has horns. That kind of dog has got to have little horns." *Photograph by Paula Lerner*

gible. For example, mathematicians share an intellectual culture that is very flexible in giving up assignments of symbols to objects and adopting others as an expedient. This flexibility is probably due to the neutral names used, such as *x*. On the other hand, disciplines that deal with tangible material and employ evocative words are dogged by energetic semantic argument. A tree feels so real that it is hard for an ecologist to remember that it is just an arbitrary assertion. To be fair to ecology, there have been many semantic arguments that have been resolved by the invention of a new term for one of the meanings in a confused literature. For instance, the new term *interference* indicates a particular type of competition. The luxury uptake of nutrients, beyond what is needed for growth, is interference, for it is an active program that aims to deny resources to the competitor. By contrast, generalized competition is more passive. It is a consequence of merely consuming resources for growth that the competitor could have used. Refining the meaning of the term *competition* by introducing types of competition clarified a semantic argument that might have been resolved more quickly in a less tendentious field, such as mathematics. Studying tangibles gives the illusion that the scientist deals with the material world directly and therefore has the one possible, and necessarily correct, definition.

A different type of semantic problem arises when a pair of definitions are locked together but are mutually exclusive. An example is the Zen question, "What is the sound of one hand clapping?" *One hand* and *clapping* are incompatible. Hidden in the conundrum is a level-of-analysis problem; "one hand" locks the discussion to a lower level than "clapping." Internal inconsistency can take many forms.

Hierarchy theory's approach is to identify problems that rest on conflicts of definition that generate internal inconsistency. By clarifying the observation process, conflicting levels of analysis may begin to be teased apart. "Chicken/egg" problems can be recast to avoid the question of which came first, but only by acknowledging multiple levels. Note that looking more closely at egg laying is of no help, for the solution is in debunking the question. As a theory of the observer's role in observation, hierarchy theory often operates outside of the realm of empirical testing. This is not to deny that insights gained from semantic arguments can sometimes help to structure further empirical tests.

In addition to definitional and level-of-analysis problems, yet another class of error is to assert that answers to questions are found only by empirically reducing from a higher level to a lower level of analysis. For

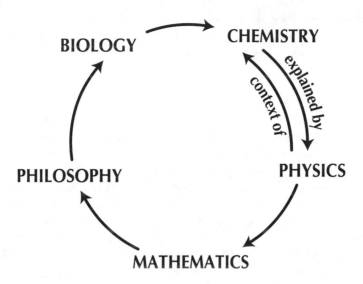

FIGURE 1.12 The circle of science
While reductionists are happy reducing psychology to biology, biology to
chemistry, and chemistry to physics, and seeing physics as applied mathematics,
the reduction closes to a circle in that the study of the origins of mathematical
thought sits squarely in psychology. One direction gives explanation, whereas
the other direction gives context and meaning.

example, it seems intuitively obvious that to explain biology means to
reduce it to chemistry; to explain chemistry is to reduce it to physics; to
explain physics is to reduce it to mathematics. But then comes the
quandary of an explanation of mathematics in terms of psychology. The
development of mathematical understanding arises as a subdiscipline
inside cognitive development, and the psychology of mathematical
insight is explained in the psychological subdiscipline of cognition (figure 1.12). While much of modern science is reductionist, that line of reasoning forgets the value of the upper-level context. Meaning, and
explaining the "why" of a phenomenon, come from the context. The
lower-level mechanics, the "how" of the phenomenon, have nothing to
say about "why."

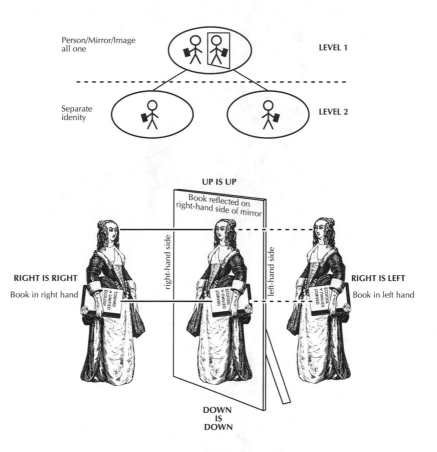

FIGURE 1.13 The mirror switches you left/right but not up/down. There is no conflict, because up and down are different logical types, each belonging to a particular level and analysis. Analyze the person and the image to give each its own identity (a lower-level analysis), and there is a switch wherein the book held in the woman's right hand is in the left hand of the autonomous image. However, analyze the reflection at a higher level of analysis, where the person, mirror, and reflection are all parts of one system, and there is no switch. The top of the mirror reflects what is pointing upward on the person, the bottom reflects what is down, and the right side of the mirror reflects the book that is in the right hand of the woman. Left and right are possessed by the woman in herself, whereas up and down do not just refer to her but must include her context. *Drawn by Kandis Elliot*

Port wing

PORT

BOW

STARBOARD

STERN

Academic disagreement often revolves around questions of the appropriate level of description, or the most useful set of definitions. Sometimes each party to an argument is unaware that the other is using a different reference, leaving no open channel for communication. Here hierarchy theory's contribution is to lay out the differences that follow from invoking this as opposed to that level. Once these subtle changes in meaning are clear, then informed choices can be made. The key words here are "appropriate" level and "useful" definitions. These terms are a matter of purpose, and that we all decide for ourselves.

Hierarchy theory helps scientists and scholars distinguish between mere semantic differences and situations where data could indeed indicate which is superior between two truly competitive theories. Two theories are truly competitive when they use the same definitions and operate on the same level of analysis. Then more data can help. However, two theories appear competitive but are in fact working at cross-purposes when they use the same words perhaps to mean different things, or are focused on different levels of analysis. Then data cannot indicate which theory is superior.

Depending on the question posed, a single material system may require a different level of explanation for each scale and type of measurement. Both scale of observation and type of measurement are chosen, perhaps implicitly, by the observer. When more than one measurement scale is used, different observers may offer explanations that are strikingly different, indicating conflict. Often the apparent conflict disappears when the consequences of implicit scaling decisions become explicit. As an example, consider the history of theories of color vision.

In the nineteenth century the Trichromatic theory of color vision rested on the fact that when red, blue, and yellow lights are reflected together, they give the perception of white. Herman von Helmholz proposed that there must be three sorts of receptors in the human sensory system, one for each color. A second fact about color mixing is that for any color there is another color of exactly the right shade that will mix to

FIGURE 1.14 Using left and right to organize navigation, some mariners might mean the other captain's right hand as the referent, whereas some might use their own left and right. Port and starboard make sure that everyone is using the logical type, given the problem at hand—ships moving relative to each other. *Drawn by Kandis Elliot*

give the perception of white. Based on this complementary color mixing, Ewald Hering proposed the Opponent Process theory of color vision. According to the Opponent Process theory, color vision is a matter of opposing signals that tug at perception to give the particular hue that is experienced. When the two sides of the tug-of-war are equally matched, the net effect is to cancel the perception of either side, resulting in the perception "seeing white." On the face of it the Trichromatic and Opponent Process theories appear directly competitive. If one is correct, then it must follow that the other is incorrect. The theories even differ in the number of fundamental entities invoked—two opposing signals versus three receptors—let alone differences in the possible mode of action.

It emerges that there are physical neurological structures that correspond to both theories. Helmholz was right! There are three types of color receptors, or cone cells, in the retina that respond to either short-, medium-, or long-wave electromagnetic frequencies to give the perception of red, blue, or yellow light. But Hering was also right! At deeper levels of processing, centers in the thalamus and occipital lobe of the cortex combine signals from several retinal cone cells. These higher brain centers perform the functions of the Opponent Process theory. The conflict between the two theories is overcome by invoking different levels of signal processing. The old way of phrasing the question, "Which one is right?" became obsolete as models of visual information processing expanded to include multiple levels.

Several levels of explanation may similarly be required for an abstract system. In conceptual schemes, changes in level amount to changes in logical types. The notion of logical types was first raised by Alfred North Whitehead and Bertrand Russell in *Principia Mathematica* early this century, but a modern systems account of logical typing can be found in Gregory Bateson's *Mind and Nature* (figure 1.13). In hierarchical classification schemes, such as animal-carnivore-dog-Spice, each branch of the hierarchy represents a change of logical type. Our intelligent and affectionate rottweiler, Spice, is a different logical type from dog in general, which is a different logical type from carnivore, which is a different logical type from animal. However, carnivore and herbivore are of the same

---

FIGURE 1.15 The four panels shade what is the system, and leave its environment unshaded. It is the same material system, but the explanation for its working depends on how system is distinguished from environment. *Drawn by Kandis Elliot*

THE AUSTRIAN TRAGEDY.

ARCHDUKE FRANCIS FERDINAND.

DUCHESS OF HOHENBERG.

AUSTRIAN HEIR AND HIS WIFE MURDERED.

SHOT IN BOSNIAN TOWN.

A STUDENT'S POLITICAL CRIME

BOMB THROWN EARLIER IN THE DAY.

THE EMPEROR'S GRIEF

The Austro-Hungarian Heir-Presumptive, the Archduke Francis Ferdinand, and his wife, the Duchess of Hohenberg, were assassinated yesterday morning at Serajevo, the capital of Bosnia. The actual assassin is described as a high school student, who fired bullets at his victims, after a bomb thrown at them earlier in the day had been deflected by the Archduke and had exploded near the second carriage of the royal procession.

The assassin is a compatriot from Trebinje, and it is stated that another is now in the hands of the police.

The news of the crime, which had been expected for some time, caused the most painful impression at Vienna and throughout the Empire.

The Emperor, who was at Ischl, was greatly affected.

SCENE OF THE MURDER.

(FROM OUR SPECIAL CORRESPONDENT.)

SERAJEVO, June 28, 9.30 P.M.

To-day at 9.50 a.m. the Imperial train conveying the Archduke Francis Ferdinand and his Consort arrived here from Ilidže. After inspecting the troops on the Filipovići parade ground the august visitors drove in a motor-car along the auction road and the Appel Quay to the Town Hall.

The crowd was enthusiastic in its greeting, but as the car was being driven along the Appel Quay, just before reaching the Cumuria Bridge. An Aide-de-Camp

---

OFFICIAL REPORT OF THE CRIME.

POPULAR INDIGNATION.

ATTEMPT TO LYNCH THE ASSASSIN.

(FROM OUR OWN CORRESPONDENT.)

VIENNA, June 28

The head of the assassin is described on the reigning House of Austro-Hungary and added yet another to the list of terrible tragedies which have befallen the House of Hapsburg.

The Archduke and his Consort were paying an official visit to Serajevo on the close of the manoeuvres of the 15th and 16th Army Corps, which had been taking place in the neighbourhood. They arrived at the capital early this morning from the little watering resort of Ilidže, which had been their head-quarters during the last few days, and to which they were to have returned this afternoon before leaving for Vienna in the evening.

The account of the crime given by the Austrian Official Telegraph Agency, which is being distributed by the newspapers as an extra edition here in the Vienna streets this afternoon, is as follows:—

"As his Imperial and Royal Highness the Archduke Francis Ferdinand, with his Consort, was proceeding this morning to a reception in the Town Hall, a bomb was hurled at his motor-car. His Imperial and Royal Highness warded it off with his arm. The bomb, which exploded after the Archduke's motor-car had passed, Count Boos Waldeck and the Aide-de-Camp of the Governor, Lieutenant-Colonel Morizzi, who were in the next car, were slightly wounded. Of the public in persons were injured, some slightly, some severely. His ... a typographer named Cabrinović from Trebinje.

"After the reception in the Town Hall, the Archduke continued, with his Consort, his drive through the town. A student named Princip, belonging to the higher class in the public school (gymnasium), a native of Grahovo, fired several shots at the motor-car with a Browning pistol. The Archduke was hit in the face and his Consort was wounded by a shot in the abdomen.

"The Archduke and the Duchess were taken to the Konak (Governor's Palace), where they succumbed to their injuries.

"Princip was arrested. Both he and the man who threw the bomb were almost lynched by the infuriated crowd."

THE BOSNIAN CAPITAL

MEDLEY OF RACES AND COSTUMES

Serajevo has a population of about 41,000 of whom 18,000 are Mahomedans and 4,000 Jews. It is the residence of a Roman Catholic Archbishop and a Greek Metropolitan, and has an Austrian garrison. One can still see at a short distance the characteristic features of Bosnia and Herzegovina, that Serajevo

---

AUSTRIAN EMPEROR AND THE NEWS.

A TRAGIC CONTRAST:

RETURN TO-DAY FROM ISCHL.

(FROM OUR OWN CORRESPONDENT.)

VIENNA, June 28.

The Emperor, who went to Ischl yesterday for his intended summer residence, received news immediately of the double assassination. The first person to express condolence to his Majesty was the Duke of Cumberland, who drove over in his carriage from his residence at Gmunden.

The Emperor is expected to be back in Vienna at 6 o'clock to-morrow morning. Three children of the Archduke are staying at the Castle of Chlumetz in Bohemia, whither the Duchess of Hohenberg's brother-in-law, Count Waldburg, at once proceeded to break the news to them.

In Vienna the flag are being flown at half-mast and flags on the palace and the foreign Embassies, but there is little trace of public excitement.

VIENNA, JUNE 27.

The Emperor Francis Joseph left here at 8.10 this morning for Ischl, where he intends spending the summer. The route to the railway station was gaily decorated and lined by an enormous crowd, which heartily cheered his Majesty.

... the Emperor was received by the Burgomaster and the members of the Municipal Council. The Burgomaster expressed the gratitude ...

KING GEORGE'S SORROW.

A WEEK OF COURT MOURNING.

The King commands that the Court shall wear mourning for one week for His Imperial and Royal Highness the late Archduke Francis Ferdinand, K.G., from to-morrow, to-day, Sunday, the 28th inst., and on Sunday, July 5, the Court to go out of mourning.

The Lord Chamberlain is commanded by the King to announce that the week of mourning arranged to take place this evening is postponed on account of the lamentable death of the Archduke Franz Ferdinand of Austria-Este, K.G., and the Duchess of Hohenberg.

The news of the Royal Family were inexpressibly shocked yesterday afternoon by the receipt of the news of the assassination.

During the evening Councillor Sir Charles Cust, the equerry-in-Waiting to the King, called at the Austrian Embassy on behalf of his Majesty to express the deep sympathy and sense of the Queen to be forwarded to the Austrian Court.

The shock which the news caused to the Royal Family was the more profound because all who knew and appreciated his person on account of the qualities of the Archduke and the Duchess had so recently been their guests.

TRAGEDY OF A ROYAL HOUSE.

THE AFFLICTED EMPEROR.

SUCCESSIVE STROKES OF FATE.

"There I am to be spared nothing" was the cry of anguish wrung from the Emperor Francis Joseph by the murder of his wife. And fate seems again to answer "nothing."

After all the awful blows that have fallen upon him as a Sovereign and as a man, it has finally happed by all who have watched the vicissitudes of his troubled life that the head of the House of Hapsburg, the oldest of European Sovereigns, might have been spared this last crowning sorrow. He might have been left, after a reign of 66 years, as he had lately recovered from a long and dangerous illness, but no one could foresee that his fifty upon him once more. And the stroke is the more terrible because it is of the same kind as those which have so often visited him. Brother, son, and wife were torn from him, one after the

---

LINER ON THE ROCKS OFF DONEGAL.

ALL PASSENGERS SAFE.

MESSAGE FROM THE CAPTAIN.

The Anchor liner California, which was on a voyage from New York to Glasgow, ran on the rocks at Tory Island, off the Donegal coast, yesterday afternoon. It is understood that the had about 1,000 passengers on board, but the latest news received early this morning is that, although the position of the liner is serious, no lives have been lost. The accident occurred during a thick fog.

In reply to a wireless message from The Times, Captain Crowley, of the California, at 2.20 this morning sent the following details of the accident:—

California ran aground, Tory Island, in fog about half mile from lighthouse. Did not hear foghorn blowing. Quite safe. No danger, no lives lost. Am not aware our accurate. Casualties standing by for transfer passengers.

COVERLEY, Master

The California went on the rocks with such force that the bow part of her broke away. Wholly strewn in mid ten feet holds soon filled with water. She is in five fathoms of water forward and seven fathoms aft. Another steamer in attendance ready to take off the passengers. Soon has been received in Londonderry that the landing of the Irish passengers may be expected before noon to-day.

The news which the coastguard were caught by the Malin Head wireless station and the entire torpedo boat destroyer flotilla on duty off the coast, looking for survivors, was sent off to all the destroyers from the cruiser fleet in Lough Swilly to hurry with all speed to the scene of the accident. Subsequently, orders were received by all the destroyers and patrol steamers on the coast from Bangor to Buncree, Co. Donegal, to keep office open all night. By 11 o'clock six destroyers were making for Tory Island.

OPERA.

logical type. Carnivores and herbivores differ from each other within a common framework. In both cases the animal is evaluated using the same criterion: type of diet.

Explanations that apply to an individual dog do not necessarily apply to all dogs, or to all carnivores for that matter. This may seem obvious, but correcting errors of mistaken logical types is a fundamental process in children's learning. Young children often find it confusing that their home is simultaneously located in a city, a state, and a country. Typically they will oscillate between asserting that one or the other of these locations is correct, while denying that all three are simultaneously possible.

In adult life, potential logical typing conflicts may be codified in special vocabularies. For example, port and starboard in nautical terminology are not just to sound salty but are there to avoid collisions. "The right-of-way is to the ship on the right!" Well, all right, but whose right did you have in mind? Port and starboard make sure that everyone is using the same level of analysis, the level of the boat defined by its designed direction of motion. The space shuttle usually flies in orbit upside-down and backward, but starboard is always to the right when you face the pointed end while standing on deck. The starboard engine is always the one that would be over your right shoulder if you were sitting in the captain's chair, even if in fact you are facing the cargo bay from the cockpit, and the shuttle is going backward with the Earth above you.

Tolerance for multiple levels of explanation is at odds with reductionist assertions that, for a given material system, there is one singular level at which resides the proper explanation, the real mechanism. Take a material system that is just a thing in a box, its environment. The thing behaves and influences its environment inside the box. Perhaps it is a fish in an aquarium. At one level appropriate measurement may uncover a mechanism that explains how the behavior of the fish influences its aquarium environment. For example, models of laminar flow may capture the influence of fish swimming on water displacement (figure 1.15).

Now slightly expand the definition of the thing, so that it includes some of what was defined as environment. In the aquarium we might

FIGURE 1.16 The Times of London for June 29, 1914. It is clear from the tone and substance of the articles that nobody had any idea what was about to unfold. (The photographs have been re-created from those of the period. The articles have been reset in the original typeface, although this is an amalgam of two of the original pages.) *Typesetting and computer graphics by Kandis Elliot*

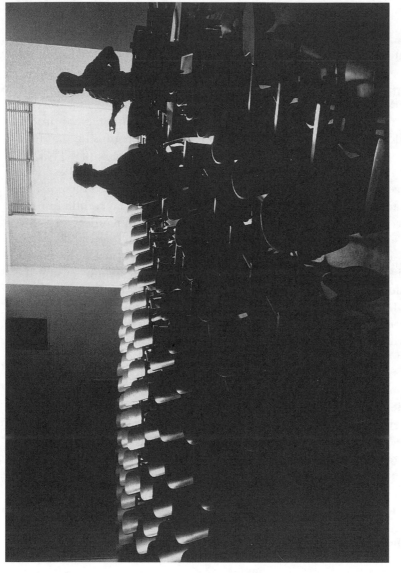

FIGURE 1.17 "He had just given the students a brilliant solution to a problem they did not know they had." *Photograph by Paula Lerner.*

expand the fish to include the green plants, and so the "thing" becomes biota, not just the animal. Behavior of the newly prescribed situation will require a different set of explanations. Note that we are still dealing with the same stuff but have just reallocated some of it between foreground and background. The material system is precisely the same, and yet any singular, proper level for the mechanism evaporates with a subjective decision to redefine the focal entities within the system. Since the material system is the same, and the only difference is the scientist's decision, no amount of probing the material system will have anything to say about which explanation is fundamentally correct, or which is the true mechanism. Notice how each subtle change of nuance in definitions changes what is considered an appropriate, or proper, explanation. Each "best" explanation is completely local to the question posed.

The crucial point here is that different modes of explanation are usually not in conflict. Each answer is relative to its own level of analysis. The choice of one answer as superior to the other rests on the nature of the question asked. Do we want to know about hydrodynamics, or is our question whether the object of interest exhibits photosynthesis and respiration in what is now the unified behavior of all the living material in the aquarium?

These same issues arise outside the natural sciences. For example, one could ask which precipitating factor triggered World War I. If we want to know if the murder of the Austrian archduke was a significant cause, is evidence of the explosive power of the charge critical? At one level the archduke's death was blamed on the German government, probably unfairly, and that caused the declaration of war. At another level, if it was not the archduke's murder, it would have been something else. The choice of question determines not only the answer but also the appropriate mode of explanation. The material system has nothing to say about which mode of explanation is appropriate. That is to be found in the values of the questioner.

Conclusion

Hierarchy theory is a theory of the role of the observer and the process of observation in scientific discourse. It is a theory of the nature of complex questions that focuses on observation as the interface between perception and learning. Hierarchy theory has roots in a number of disciplines, including psychology and philosophy. It differs from a ritualistic or mechanical approach to complex problems in that instead of achieving predictability by gaining tight control over a small number of factors, hierarchy theory's method is to expand the problem space to include the observer as well as the observed.

Hierarchy theory examines closely issues of definition, measurement scale, and purpose in scientific models. Often extremely subtle changes in nuance, usually due to a shift in logical type, causes two theories to appear conflicting when in fact they operate at different levels of analysis. Hierarchy theory offers an advantage in disputes where collecting more data, that is to say looking again more closely, does not help.

In this book we will take the reader through a series of examples and thought experiments that challenge realism and reductionist science and offer hierarchy theory as an alternative. Hierarchy theory is a holistic theory, and the goal of this book is to make holistic thinking available to everyone. A colleague once told the young Assistant Professor Allen that he had just given the students a brilliant solution to a problem they did not know they had. As a result, his lecture had been, on balance, a failure. Wishing to avoid making the same mistake again, we will have to spend some time in this book giving a problem to those satisfied with reductionist orthodoxy. There is a problem with rude, realist determinism. This book assaults that position but offers a constructive alternative that has broad application and intuitive appeal.

2

Levels of Analysis as a Challenge to Realism

Levels and the Nature of Complexity

Many disciplines assert that their material is complex. Ecological communities, cells, and social institutions are all viewed as such by scholars in their respective fields. Despite this agreement among the various practitioners, there is remarkably little consideration of what it means to say that something is complex. If one could identify what is required for something to be complex, then one might be able to deal more effectively with it. What is it that makes some things simple, and what happens differently to make other things complex?

Observation is the stock and trade of science, and the very act of observing necessarily employs a point of view. By committing observation to any one particular perspective, all other perspectives are rejected until there is a change of commitment. One can make anything simple by identifying it by only a small number of commensurately scaled characteristics, and ignoring all others. Thus by describing a system in such a way as to make it appear simple, for the moment, one is not treating it as complex. However, that does not invalidate the many other views that would make the same material system indeed appear complex. For a given material system, both simple and complex descriptions have their place, but a move from one to the other requires a change of perspective.

The Strategy of a Hierarchical Approach

Hierarchy theory is a theory of the observer's role in any formal study of complex systems. We defined a complex system as one in which fine details are linked to large outcomes. In order to describe adequately a complex system, several levels need to be addressed simultaneously. Levels may be ordered according to the scale at which each operates, and scale of observation is fixed by the measurement protocol. Complexity therefore involves relating structures and processes that are observed at different scales. Reductionism deals with compexity by narrowing focus on system parts so closely that they are forced to appear simple, at least when disaggregated. By focusing on issues of *scale, levels of organization, levels of observation, levels of explanation,* and relationships between these levels, hierarchy theory offers an alternative to mechanical, reductionist approaches to complex systems.

In simple systems the effects of low-level details can usually be ignored. Fine details reside at low levels of organization. The effects of behavior of these details in simple systems usually disappear, so that no matter what happens at the low level of fine details, everything important coming from lower levels turns out to be some sort of average condition. In simple systems, if there are any effects of differences at the level of fine details, such that a value other than average emerges, its effects usually attenuate to nothing quickly. By contrast, in systems that invite the researcher to call them complex, the effects of fine details amplify so that small differences end up having large effects. In this way, low-level details in complex systems can exert an influence over high levels and affect the behavior of the whole system.

A simple problem is adequately described using only one level of analysis as the central explanation of the observed behavior of the system. If the problem is simple, the only important levels are the level of the phenomenon, and the single, homogeneous level of explanation below it. Since all levels below the explanatory level exhibit average behavior, their behavior can be ignored, and a single value can represent each significant low-level state. For instance, there are differences between atoms of iron at room temperature, but for many purposes they interact to produce a consistent character of homogeneous solidity. Furthermore, levels above the phenomenal level exhibit no behavior, so they too can be taken for granted. Here, for instance, the atmosphere can be taken as a constant context for living systems, its oxygen content not having changed very much for well over a billion years. By contrast, in a complex problem, the effects of deep lower-level conditions do not disappear into an average condition. Instead, details of lower-level dynamics influence upper-level behavior to make it rich and difficult to

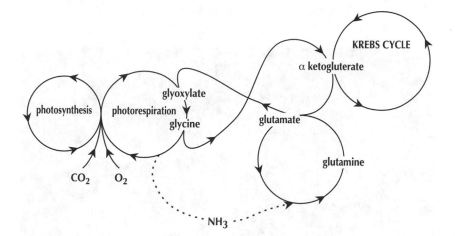

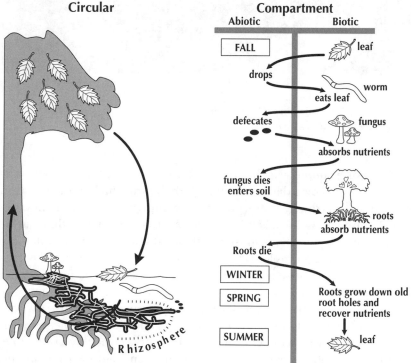

FIGURE 2.1 Both ecological nutrient cycles and biochemical cycles are investigated by introducing radioactive tracers into the system and finding out how long the radioactivity takes to occur elsewhere in the system.

FIGURE 2.2   "Organized in a lattice configuration."
Photograph by Paula Lerner

predict. For example, which individuals will rise as political figures cannot be predicted over decades, although their individual quirks can seem to change everything. Predictability in complex systems is achievable only if many levels are taken into account. Deep low levels that can be ignored in simple systems must be taken into account in complex systems. A problem is complex when an explanation of its associated behavior requires several disparate levels to be addressed simultaneously.

A system is defined as hierarchical if it can be described as composed of stable, observable subunits unified by a superordinate relation. Hierarchical subunits are not necessarily tangible. For example, nutrient cycles in ecology or biochemical cycles in cells are reliably observable, even though nobody has ever literally seen them. Both ecosystem and cell cycles can be detected by the movement of radioactive tracers (figure 2.1). Indirect observation of intangible elements is just as valid as direct observation. For purposes of a hierarchical analysis, any reliable measurement may well be sufficient.

An example of a system that can be described as a hierarchical system is a chair. Molecules can be considered stable subunits that form the level below the whole chair. The molecules are organized in a lattice configuration, which is the superordinate relation. This relation unifies the parts into a whole.

For most purposes a chair can be described adequately as a simple system. The behavior of the lower-level molecules—for example, their rate of vibration—dissipates in the average condition of solidity. Solidity is an upper-level emergent property that is not itself a feature of the individual molecules at the lower level. The philosopher John Searle is wonderfully clear in demystifying the notion of emergent properties in upper-level aggregates. The chair might need to be considered a complex hierarchical system if the superordinate relation, the lattice, begins to weaken and the chair falls away below the person sitting on it. If the question is the load that one might expect the chair to take before breaking, it can probably still be treated as a simple system. On the other hand, if the question is about where in detail, and in what manner, the chair will disintegrate, then it probably requires treating as a complex system.

Predicting which molecules are going to liberate themselves, and so which section of the chair is going to break first, is a complex problem. In this case the behavior of the lower level cannot be ignored. Fine-grain details reorganize the structures of the upper level, and someone sitting on the chair ends up on the floor. In both the stable and the collapsing chair we are dealing with the same hierarchical system composed of the same subunits, but the stable chair is simple, because it can be represented by an average chair, whereas the breaking chair is complex, if one needs to predict the details of its demise. The difference is that in the complex system, the details of lower-level behavior have a profound effect on the upper level.

Notice that units and levels should not be interpreted as features of the external world, existing independently of an observer's criteria for delimiting the system. The concept of level is relative to the point of view taken by an observer. Since complexity comes from the relationships between levels, it is to be expected that complexity is not a feature of the external world. Complexity does not exist independently of an observer's questions. Instead, complexity is the product of asking questions in a certain way.

With regard to the chair, if its reliable use is as something on which one might sit, then the chair is a simple system. If we load the chair until

FIGURE 2.3 "Simple questions can be answered by relying on the performance of an average chair of the type being tested." *Photograph by Paula Lerner*

it breaks, questions about how much of a load it can stand may also be simple. Experiments on similar chairs will suggest that perhaps three hundred pounds can be supported, but three hundred and fifty pounds is questionable. However, questions as to the manner of the disintegration of particular chairs will almost certainly be complex, since slight imperfections in the details of the lattice structure of the chair, details unique to the chair in question, will be the only reliable way to answer the question. When details are important, it is because they amplify their influence through positive feedback. It is usually very difficult to determine which breaking bond will cause others to break and which will remain isolated events. The simple questions can be answered by relying on the performance of an average chair of the type being tested, but the complex question requires relating lower-level details of the specific chair's fine structure to the upper-level context, the chair's integrity through a chain reaction. If a problem requires taking into account both fine-grain details and aggregate or rich emergent behavior, then multiple levels of organization are required for a solution. However, it is the model embedded in the question, not the material system itself, that is complex.

The Process of Scientific Investigation

So far so good. Hierarchy theory sounds like a reasonable idea, but how do you begin to apply it? The starting point for identifying subunits, relations, and levels is to make explicit those aspects of an observation which come from the observer and those which come from the observed (figure 2.4). In this chapter we outline a five-step process of scientific investigation. At each step the contributions of the observer are separated from those of the observed. Keep in mind that these steps represent an ideal sequence. We do not imply that working through the five steps below is how scientists do in fact conduct the mechanics of their investigations day to day. They jump around, invoking the various activities as they suit the particulars of the moment.

The purpose of this chapter is to cultivate an abstract model of observation. Let us start with the observer. There are five junctures at which an observer's decisions are crucial to structuring an observation. These are: 1) posing a question; 2) defining entities or units; 3) choosing measurements; 4) noticing phenomena; and 5) evaluating models. Each step invites errors of logical typing and naive realism. In order to avoid such

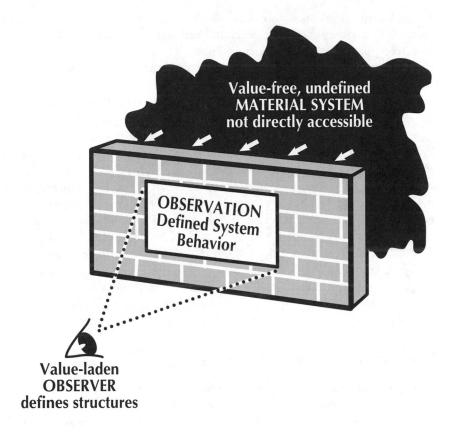

FIGURE 2.4 Observation never involves access to the value-free material system, for scientists have access to only defined system behavior, which must involve the value-laden observer. There is no such thing as objectivity in science—only cognizance of biases, if you are lucky.

errors, we will divide observation into a sequence of alternating contributions from the observer and the observed.

Decision 1: Posing a Question

There is a difference between sensing and observing. Sensation can be defined as mere physiological activation of the central nervous system. By this model, it is a passive response to impinging environmental energy. Observing, on the other hand, is an active process. To make an observation one must search but cannot begin without an idea of what it is that one might find. Accordingly, the process of observation begins when a question is posed. As soon as a question is formulated, certain constraints immediately follow. The question directs and structures the observation process by posing hypothetical entities and their roles.

Decision 2: Defining Entities or Units

Before an observation can be made there must be a "thing" to observe. Which entities are going to play a role in the investigation must be chosen by the observer. In disciplines that focus on tangibles, such as trees or organisms, it may feel as though the distinction between "thing" and "not-thing" is given by the world. However, consider intangible entities like the money supply, or personality. Are these, too, given by the world, or are intangible entities the product of an entirely different procedure? Does the process of observation itself vary by whether a discipline deals with tangibles or intangibles?

The answer to these questions is an emphatic no. Both tangible and intangible entities derive from the same process. In neither case are they given by the world. Instead, the boundary between "thing" and "not-thing" is found in the observer's criteria, or definitions. Sensation gives raw, uninterpreted experience. Perception, on the other hand, is of things *as* "things," that is, as members of categories. Cognition bridges between sensation and perception by linking the data from sensation to expectations, categories, and conceptual schemes. Cognitive interpretation of sensory input allows meaningful perceptual experiences. Criteria shape sensations into the experiences of things *as* "things."

Definitional criteria highlight some aspects of experience, and lead us to ignore the rest as continuous background. Observation differentiates between figure and ground. Figure is that part of the observation field that is treated as significant, and ground is everything else. According to this model, the boundary between figure and ground is not given by the external world. Both figure and ground are the product of an observer's assertions, questions, values, prior beliefs, and expectations. Before criteria differentiate experience into figure and ground, there is no *thing* to behave. This set of distinctions is useful because one common error in scientific arguments is to argue over which entities are real or true. By

FIGURE 2.5 "When one is on the surface of the sea, many things of interest are at a distance below." *Photograph by Paula Lerner*

regarding entities as the product of an observer's decisions, that unproductive line of argument is blunted at the outset.

Perception and observation are indirect processes that rest on sensation and cognition as intermediaries. Realist tendencies are most likely to arise in disciplines that deal with tangibles, probably because tangibles feel so direct and intuitive that it is counterintuitive to imagine that observation is indeed indirect. Related problems arise in disciplines where the object of study is similarly scaled to ourselves or, worse yet, exactly the same as ourselves, as in psychology. In all these disciplines observation is indirect, but it is easy to forget one's tools. In fields where the objects of study operate on a vastly different scale from ourselves, such as biochemistry or astronomy, the practitioners within the field are acutely aware of the indirect nature of observation and are less prone to problems of reifying their subject matter.

In order to help avoid the error of reification, some of the examples used in this primer invoke nautical imagery. We will use the metaphor of a fishing net in several places. This is not a coincidence. It is very hard to avoid realism and concrete thinking when one is surrounded by tangibles. When one is on the surface of the sea, many things of interest are at a distance below, and one is forced to acknowledge the role of tools in making things part of experience. That is the reason that images of the sea and fishing work so well. Observation, cognition, and perception are all intermediaries between the observer and what is, in the end, observed.

Decision 3: Choosing Measurements

Once entities of interest are defined, the next issue relates to what they do and how they behave. Measurements tap behavior by giving access to changes of measured state over time. The differential between the state of the entity E at time one (E_{t1}) and entity E at time two (E_{t2}) gives behavior ($B = E_{t2} - E_{t1}$). Behavior is inferred from entity changes of state over time. Which entities and which behaviors are observed depends on the scale of the measurement protocol chosen.

So far the observer has chosen focal objects and type of measurement. Once these decisions are made, system behavior is now free to emerge. Think of the world as constant dynamical flux. In contrast to the observer, the observed is simply out there, doing what it does, without regard for either the informal observer or the formal scientist. However, once definitions and a measurement regime are in place, the observed rather than the observer is responsible for changes of state. Thus behavior is that part of the observation process that is above and beyond the observer, and therein lies the generality of scientific investigation. Two different observers, given the same definitions and measurement protocol, will observe the same behavior. Both observers isolate the same aspect of independent, external, dynamical flux.

If objectivity is defined as "free from human interference and interpretation," then the aspects of behavior that are beyond the control of the measurement protocol are objective. This objective phase of science does not last long, however, because measured behavior is immediately evaluated for its fit and significance in existing models and expectations. Mere changes of state, like mere passive physiological arousal, are uninterpreted and therefore meaningless. If observed behavior is not going to sink into undifferentiated background change, it must be highlighted. It must be given a label, a category, and be integrated into a conceptual scheme.

Decision 4: Noticing Phenomena

Once system behavior has been registered by a measurement protocol, only some changes of state will be considered interesting. The rest will be ignored. This decision process mirrors the division of experience into significant figure and incidental ground. Behavior is evaluated as interesting or not, and those behaviors deemed significant often become associated with the figure. Significant behavior in the ground might be a source of stress or disturbance. The rest becomes noise or background details and is discarded. *Phenomena* are defined as those focal behaviors deemed worthy of further investigation. Once again, the observer is active. Entities and phenomena are both the product of an observer's values and decisions as to what is worthy of further attention. Experience that either fits into or obviously violates existing cognitive schemes becomes figure, and the rest gets left behind as background.

While decisions of what is phenomenal and what is only incidental come from the observer, these decisions are not made capriciously. Significant events are those that either quietly fit into or, better yet, fill gaps in existing knowledge. Behavior is evaluated by relating it to expectations and prior knowledge. These decisions are subjective in the sense that their source is the observer, but they are not entirely independent of changes in the material system.

To illustrate the role of phenomena in science, imagine a spider that has woven an exquisite web. The spider in this example is analogous to a scientist, and the web is analogous to a network of existing beliefs or knowledge within a discipline. Willard Quine spoke of a "web of belief," and we borrow that notion here. Now consider what happens when an observation is made. An observation is like a perturbation to the web. If an edible bug lands in the web, it may temporarily disrupt the web; but not to worry. The spider is soon actively at work, carefully packaging its meal and then repairing the web. In the end there has been progress. The web is better than before for it now contains food, which was indeed the object of the enterprise in the first place. This situation is analogous to a scientist, working within a particular discipline, making a discovery. The scientist's state of knowledge, and the discipline itself, are advanced.

Now consider all the less than ideal perturbations that might affect the web. Perhaps it is not a bug but a drop of rain that temporarily destroys a portion of the web. Not all is lost, because the spider quickly

FIGURE 2.6 "Imagine a spider that has woven an exquisite web."
Photograph by Robert Mitchell

patches up the damage. Webs and belief networks always need maintaining. If the web happens to be in a place where it is regularly disrupted to no advantage, the spider may well even move to a different location. A line of inquiry that necessitates too many patches is best abandoned.

Different species of spiders respond only to insects of the right size. Very small insects are ignored and accumulate as dross in the web. Stinging insects that are too big are a threat, and are actively avoided. The spider leaves them alone until they go away. Disruption of the web by both edible bugs and raindrops is analogous to what Thomas Kuhn called "normal science" (1970). In normal functioning, the web of knowledge is continually repaired, and spiders live on the food accumulated.

Consider now what happens if not a bug or raindrop but a bird destroys the entire web. The spider's entire worldview is destroyed! It is time to build a new web altogether. This would be analogous to a paradigm shift in Kuhn's terminology. Kuhn defines a scientific *paradigm* as the dominant framework within a discipline that determines which questions are valid and which questions are not. Valid questions are of interest and worthy of further pursuit. If a question violates the paradigm, or conventional wisdom, it is often actively shunned and dismissed as nonsense. For example, in biology, evolution by natural selection is the dominant paradigm. Any questions that introduce purpose or foresight into evolution are shunned as either Lamarckian or Creationist. In fact, the tension between Darwinian and Lamarckian evolution is a holdover from the paradigm fight that established Darwin as the winner in the English-speaking world at the turn of the century. Despite the self-satisfied pronouncements of Darwinians, Darwinian and Lamarckian accounts of evolution are less than fully competitive theories. Even so, Lamarck is still heresy at the cutting edge of conventional wisdom.

When a paradigm shift occurs, the old web of belief is discarded and a new web begins to form. Modern quantum mechanics in physics is an example of a new web of knowledge that replaced the paradigm of Newtonian mechanics. When such a shift occurs, it is not because all the questions within the old paradigm are answered. Instead, the questions themselves change. Old paradigms are used for what they used to do well. Notice that we do not build bridges with quantum mechanics. Insistent proponents of old paradigms may be puzzled and irritated. Puzzled because there is no longer any interest in questions that used to be important, and irritated at all the hullabaloo over new questions that sidestep the "crucial" issues. New paradigms look like fads to adherents of old paradigms.

The spider web analogy illustrates several points. First of all, the scientist is active. Building a web or making an observation is an active strategy. It not only reflects the state of current knowledge within a discipline but is directed toward the goal of capturing ideas that will sustain both the scientist and the field. Second, not everything that touches a web is good. There is an active selective process here. Some perturbations are wrapped as prizes, some are too small or too big and are ignored, and some are so devastating that they demand reconstruction of all that existed before.

We defined phenomena as focal behaviors deemed worthy of further investigation. Phenomena are observed behaviors that have some sort of impact on a field, and are analogous to a perturbation of a web of belief. Phenomena may be kept and wrapped (a discovery), worked around (maintenance), or suppressed as anomaly and ignored. No matter what the effect, the mode of accommodation is not chosen capriciously. The scientist spider is a member of a particular species (discipline), working within a specific framework (this web of belief), and striving toward a particular goal (sustaining self and discipline). The web threads (the current state of knowledge in a discipline), the spider's health (the scientist's career), and even the strategy of web building itself (the scientific method), can all be seen as an implicit set of constraints that make decisions surrounding phenomena coherent. Decisions about phenomena are subjective in the sense that they depend on an observer's interests and decisions, but the decision making process itself is not capricious.

Let us make the process of identifying a phenomenon concrete through an example from a scientific investigation. The temporary weather cycle called El Niño brings floods and droughts from Australia to the Americas about every ten years. The pattern was well known not only to scientists but to many inhabitants of the Pacific Rim. However, only relatively recently was intermittent warming of the surface of Pacific waters recognized as the indicator of a new El Niño onset. The cause of Pacific warming is not understood. A ten-year cycle can even be seen in lakes in Northern Wisconsin, in the center of North America. About one in ten layers of sediments there is unusually thick from sediment laid down in El Niño years. Individual floods, while perhaps phenomenal in their own right, may or may not be part of the El Niño pattern. Discriminating between El Niño floods and other flood events is a matter of selecting data to match expectations. Floods also occur out of cycle, but the largest floods usually do occur during El Niño. When large floods occur, they tend to be coordinated with droughts elsewhere.

A post hoc criterion of "unusually large floods, correlated with droughts elsewhere, on a ten-year cycle" is one way to discriminate between floods that are part of the El Niño phenomenon and floods that are not. This is analogous to the spider deciding which perturbations to ignore and which to wrap as lunch. Layers of sediment, particularly thick layers, matter. The actual chemical composition of the sediment itself does not matter, and varies by region. It is ignored. The criterion of thick sediment is theory-guided but not capricious. Also notice how the criterion guides what matters in an observation and what does not. We return to a further discussion of the notion of phenomena in chapter 3, where phenomena are described in terms of rate-dependent dynamics and rate-independent significance.

Decision 5: Evaluating Models

The final juncture at which an observer's decisions structure observation is evaluating models. Although they are a critical part of science, data are not the purpose of science. Science is about predictability, and predictability derives from models. Data are limited to the special case of what happened when the measurements were made. Models, on the other hand, subsume data. Only through models can data be used to say what will happen again, before subsequent measurements are made. Data alone predict nothing.

Models may be mathematical, such as computer simulations, or word-based, such as evolution by natural selection. In collaboration with other observers of the same phenomenon, investigators evaluate the usefulness of a model and its relation to other models. Just as it was the job of the observer to decide which entities are useful and which behaviors are interesting, the observer must now decide which models are most pertinent, inclusive, and parsimonious. Criteria such as aesthetic appeal and fit within the predominant paradigm are relevant here. Many perfectly good theories fail to gain popularity simply because they are not easily seen as the next obvious extension of existing knowledge. Because they fail to address questions that come easily in the dominant paradigm, they are not assimilated into the scientific culture. In retrospect, some are recognized as having been ahead of their time. The time for some of them never comes.

Posing a question, defining entities or units, choosing measurements, noticing phenomena, and evaluating models are five distinct junctures at which an observer's decisions structure a scientific investigation. While the five decisions were presented as an ideal sequence, in practice one could bounce from any one decision to any other. One is always free to change one's question or measurement protocol at any time. The important point is that these decisions are often made implicitly, and confusion stems from either suppressing the role of the observer (realism) or not anticipating the theoretical consequences of a shift in definition or scale

---

FIGURE 2.7 "Useful definitions are those that correlate with repeated patterns." *Photograph by Paula Lerner*

(logical typing problems). Definitions of entities are conveniences chosen by observers, and as conveniences they are subject to revision. Revising definitions is just as important to scientific progress as collecting data. Useful definitions are those that correlate with repeated patterns in the observed. Once one has repeating behavioral patterns, then one is in a position to build models.

An Example of Identifying Structures and Processes

So far we have put science and the scientist in a general framework. We now work through a brief, concrete example of circumscribing entities, revising categories, and building models. We use an example outside formal scientific approaches to show that everyone, scientist or not, follows roughly the same general framework for investigating the world.

Imagine a fictional character whose sensory filters are sensitive only to carbon lattice structures. Carbon is an atom to which other atoms easily bond and is accordingly the foundation for many spatially discrete materials in our world, such as plants and animals. We name our fictional character "C," for carbon filter.

C, looking at a person sitting on a wooden chair at a wooden desk, would see a single carbon lattice occupying a continuous space. Person plus chair plus desk constitute a discrete carbon lattice figure, separated from a background of oxygen, nitrogen, and carbon dioxide haze (figure 2.8). If the person sitting at the desk stands up and walks away from the desk, a break would appear in what had been before a single, continuous space (figure moves away). This new observation could be interpreted, erroneously in this case, as a reproductive event. One large profile entity has now given way to two smaller, static entities (the table and chair) and one mobile entity (the person).

The person entity is that subset of the original carbon lattice structure characterized by bounded motion. The table and chair are two spatially discrete subsets characterized by static, continuous borders. Using bounded motion as the criterion, C labels the person "E," for entity. Here

---

FIGURE 2.8 With the figure on the chair at the desk, it would appear to someone who could see only carbon lattices that there is one coherent entity, not three. As the person stands up and walks away, two stationary objects and one mobile object emerge. *Drawn by Kandis Elliot.*

*He was beginning to quiver all over like
Lionel Barrymore.*

C began with one definition of E but had to revise that definition as subsequent observations were made. Such revision occurs all the time in science, particularly in less established disciplines or new areas of research. Of course, a person at a desk being familiar, the designation of an act of reproduction here seems absurd. However, microscopists in the middle of the last century were confronted with similar problems of isolating robust entity borders when they first saw cells and microorganisms. "Thing" versus "not-thing," and "thing A" versus "thing B," were crucial distinctions that needed to stabilize before theories of cell division, or sexual and asexual reproduction in microorganisms, could be built. Observing asexual reproduction in amoeba was probably not unlike C's observation of E's "birth." In reflecting on his university days James Thurber remarked on how, much to his professor's dismay, he could never see a plant cell down a microscope. "You were supposed to see a vivid, restless clockwork of sharply defined plant cells. 'I see what looks like a lot of milk'" (*My Life and Hard Times*, p. 89) (figure 2.9).

As C's observation of E and the desk and chair progresses, it turns out that bounded motion is useful as the defining feature of entity *qua* personhood. Bounded motion does correlate highly with repeated patterns, and no further reproductive events take place. If C were watching a movie and E was the hero, then E would reappear frequently. The point here is that the breaks between entity and nonentity come from the observer, who, given certain sensory capacities, seeks boundaries that predict recurrent and stable configurations in experience.

Now that C has defined an entity, system dynamics can begin to be harnessed. What happens between observations of E at t_1 and E at t_2? The *differences* between occurrences of E taps into dynamics and gives behavior. In this case, the behavior of E is ($B = E_{t2} - E_{t1}$). Behavior, B, is the contribution of the observed. C does not control what E does; that is E's contribution. C only chose to focus on boundaries that highlight E as distinct from all other spatially disjunct groupings (such as E plus desk plus chair).

C watches E for a while and comes up with a list of E's behaviors. C can now begin to model and try to explain what E does. Models summarize

FIGURE 2.9 James Thurber's drawing of his college teacher, enraged with Thurber's inability to see anything down the microscope. *Reproduced by permission from "University Days."*

past behavior and predict future behavior. Building predictive models is the chief goal of science, but we all use models of some form most of the time. Almost everyone has an intuitive or "folk" theory of physics, psychology, and biology.

Conclusion

Hierarchy theory is a theory of observation whose function is to clarify those arguments in science where collecting more data will not help. Five junctures were delimited at which an observer's decisions are crucial to structuring the process of scientific investigation: 1) posing a question; 2) defining entities or units; 3) choosing measurements; 4) noticing phenomena; and 5) evaluating models. These five decisions are the contribution of the observer. The contribution of the observed is system behavior. While the observer does not control system behavior, behavior does occur in the context of the observer's decisions. The world's dynamical flux is external, and beyond the observer's control. However, which behavior is observed depends on which units are chosen and how measurements are made. The generality of science comes from the fact that different observers, using the same definitions and measurement protocol, can isolate the same aspect of external, dynamical flux.

From the tidy fashion in which scientists present the results of a scientific investigation, one might get the impression that there is a fixed temporal sequence to scientific investigation. First a question is posed, and finally a model is presented. This is not, however, what usually happens. A scientific study can invoke any or all the stages in between posing a question and reporting a model. One might be assessing the utility of a model, and then for aesthetic reasons plug in a new set of asserted entities to see if they work better. Working on a problem day to day is usually messy, and progress is often not in a logical sequence. This does not imply that science is haphazard or illogical. Abandoning a logical sequence means only that science proceeds as much by other processes, such as creativity, inspiration, and guesswork, as by linear analysis.

The purpose of this chapter was to give an overview of a single level of observation. We now turn to issues of scale and levels of organization, and introduce a vocabulary of technical terms specific to hierarchy theory so that we can develop tools to deal with multiple levels of observation.

3

The Critical Dualities in Observation

In the last chapter we divided the process of observation into a five-stage sequence that begins with a question and ends with evaluating models and generating new questions. In this chapter we go further in teasing apart the contributions of the observer and the observed in scientific inquiry. A tension between the perceptual side and the material side of observation will be drawn out in two separate arenas. In one arena we deal with matters of scale, and in the other we turn to structural aspects of entities. We conclude by drawing a distinction between entities postulated before observation and those entities observed post hoc.

A firm grasp of the subtleties of scaling and structural issues is necessary for effective hierarchical analysis. Both scale and structure have observer and observed sides to them. Every observation invokes structures, and pertains to a particular scale. Thus in this chapter we unfold the dualities inherent in structure and scale. Then we will finally be in a position to take up the issue of hierarchical ordering of levels of experience. The present chapter moves toward ordering hierarchical levels by clarifying what is entailed in a single level of observation. The ordering of multiple levels of observation occurs in chapter 4, where we explore how observation cuts into different levels of experience by scaling operations.

The Observer and Observed Contribute to Scale

We defined observation as the interface between perception and cognition. Observation is structured experience. In science it is aimed toward

FIGURE 3.1 "When a figure such as a leaf on a far-off tree is seen from a distance with the naked eye, one might not even consider an individual leaf as being worth distinguishing at all." *Drawn by Kandis Elliot*

FIGURE 3.2 "On the other hand, look through a microscope, and then even a whole leaf is too large to be seen." For animals that live out their whole lives on a single leaf, it is a huge landscape. *Drawn by Kandis Elliot*

FIGURE 3.3 "One can make it appear larger or smaller by looking closely or from afar, but the elephant is intransigent and insists on being the size that it is. Look close, look from a distance, but the elephant does not change its weight." *Drawn by Kandis Elliot*

building predictive models. The point is to link the world of internal models to the dynamical flux of the external world. Meaningful experience is interpreted dynamics. Given these distinctions, the observer can be seen as active on two fronts. One is building internal cognitive models, and the other is interpreting experience.

Notice that, according to this epistemology, at no point does the observer access the external world directly (see figure 1.9). Experience is always interpreted by cognitive models, which operate at the boundary between the internal and external worlds. Knowledge about the external world tells us what is; knowledge of possibility is of what might be; whereas necessity and morality indicate what ought to be. "Is," "might be," and "ought to be" are always contained within internal models. According to this epistemology, knowledge is not a direct link between

FIGURE 3.4 "Think of catching fish with a net." *Photograph by Paula Lerner*

the observer and the world. Knowledge is of the world as we see it. How the world is in itself is the realm of ontology rather than science.

Measurement scale sets limits on the scope of what can be seen, or captured in an observation. Issues of scale enter an investigation as soon as a measurement regime is established. For example, look though a telescope, and distant figures become large and can be studied in some detail. Look with the naked eye, and one cannot see much detail in stars and planets. Similarly, when a figure such as a leaf on a far-off tree is seen from a distance with the naked eye, one might not even consider an individual leaf as being worth distinguishing at all (figure 3.1). On the other hand, look through a microscope, and then even a single whole leaf is too large to be seen (figure 3.2). Measurement protocol defines the scope of observation by establishing temporal and spatial limits. Inevitably, there are limits on the smallest and largest entities that may be captured in a data set.

Some aspects of scale are clearly associated with the observed rather than the observer. No matter what the observer does, there is nothing that can be done to change the material size of an object in the environment. Take, for example, an elephant. One can make it appear larger or smaller by looking closely or from afar, but the elephant is intransigent and insists on being the size that it is. Look close, look from a distance, but the elephant does not change its weight (figure 3.3). In order to be effective, a measurement protocol must accommodate to what is out there.

Material systems have immutable scalar properties, but this does not mean that the material world fixes the scale of observation. The material world stubbornly retains its scalar properties, but scale of observation, like criteria for foreground and background, comes from observer decisions. Even Procrustes could not stretch a water molecule to fit his bed and have it still be water. Looking for an elephant with a microscope means that many attributes of the elephant will not be seen, while other attributes will be enhanced. But it is the observer's choice of measurement, not the elephant, that is responsible for the scaling effect of observation.

Observation Protocols

Measurement, Grain, and Extent

As a metaphor for how measurement sets spatial and temporal limits on observation, think of catching fish with a net. The size of the holes in the

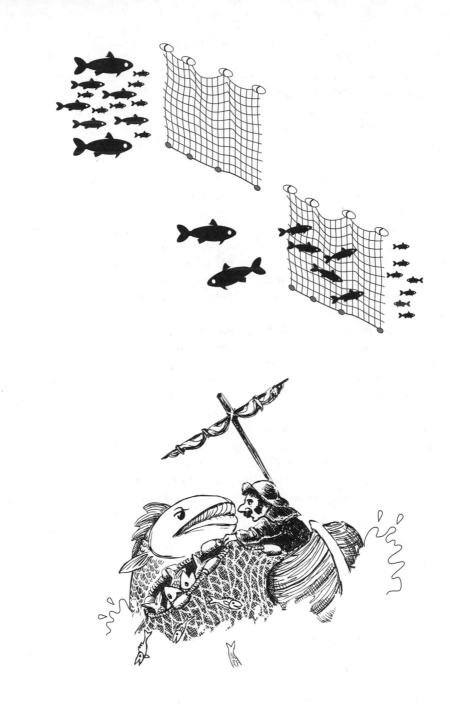

net allow some entities, such as water molecules and small fish, to pass through unaffected. Others, such as medium-sized fish, are captured, but some objects are simply too large. Whales will not fit. It will take a different strategy if one wants to capture entities that are either too large or too small for a fishing-net sampling method.

We make measurements to find differences. The fish caught in the net are different from one another, and they are also different from their watery background, which drains away as the net is pulled out of the water. Anything that is too big or too small to be part of the catch is not seen and becomes undifferentiated background. The important distinction here is size.

If fish were scientists they could readily study octopi and lobsters, but they would have difficulty discovering whales; worse than that, two of the last things they would discover are water and the ocean. Water and the ocean both form the undifferentiated context for fish, although their implicit influences are felt from the lower and upper levels respectively. In the history of science, organisms were well described inside an elaborate taxonomic framework long before atoms were given account in a coherent theory. Much as fish might discover the ocean later rather than sooner, only in this century has the redshift been used by humans to identify that the universe at large appears to be expanding and that the farthest objects seen in the night sky pertain to happenings billions of years ago. In the example of the net and the fish, the upper- and lower-level context are participating members of the oceanic ecosystem, even if they are not caught in the net. All observation has temporal and spatial limits, and occurs within an unseen, undifferentiated context that may, nonetheless, be causally relevant to the level under study. Context is important even though it is not part of the measurement directly.

In hierarchy theory there are technical terms for limits on the smallest and largest entities accessible. The threshold between the smallest things captured and those that slip through the net of observation unrecorded is the *grain* of an observation. Grain is analogous to the size of the holes in the fishing net. As water molecules pass out of the holes

---

FIGURE 3.5 The size of the holes in the net allow some entities, such as water molecules and small fish, to pass through unaffected. Others, such as medium-sized fish, are captured, but some objects are simply too large. Whales will not fit." *Drawn by Kandis Elliot*

FIGURE 3.6 "Those that are simply too big." *Drawn by Kandis Elliot*

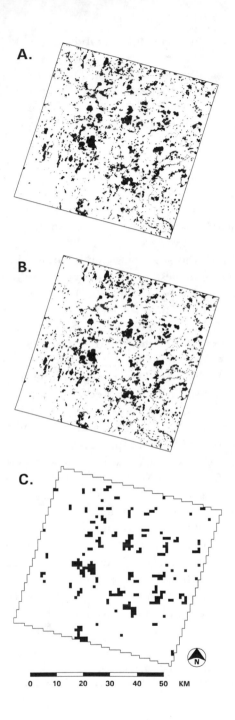

A.

B.

C.

0   10   20   30   40   50   KM

millions at a time, no distinction is made between any two molecules. They are finer than the grain of the data collection protocol, and fall between points of observation. They are not identifiable because they slip through the distinctions made by the measurement method.

Now consider the other end of the sampling protocol. Here the threshold is between the largest entities that the net can capture and those that are simply too big. This upper threshold is called the *extent* of an observation protocol. Extent is the span of the sample and is analogous to the size of the net, the area fished, and the entire period of the fishing expedition. Anything that is too large to fit into the net, outside of the area fished, or absent during the time of the expedition will be excluded from the catch. It pertains to something that is too large to be distinguished or identified, even though its influence may still be present. The *data set* is analogous to the yield of the fishing expedition. Grain and extent limits are important because they fix the scale of experience.

Scale may be temporal, spatial, or both. A large-scale study covers either a long time period or a broad spatial expanse. A small scale study is narrowly focused in space or time. Because scale can be either temporal or spatial, and scale is defined by grain and extent, it follows that both grain and extent can be either temporal or spatial, and are often both. Take, for example, a digital sound track. A sound track has no obvious spatial component, but it does have an obvious temporal grain and extent. The grain is the time period represented by each unit of digital information, and the extent is the length of the entire recording. Thus a digital sound track is principally a temporal matter. By contrast, a single photographic image of the surface of the Earth seen from space has no significant temporal component. The image is captured in all its parts essentially instantaneously. The extent is the span across the entire image, and the grain is the size of each patch of ground that is distinguishable from the next (figure 3.7).

FIGURE 3.7 Three remotely sensed images of the same lakeland area in Wisconsin. In A, much detail can be seen because the grain is in meters. In B, a different data-collecting protocol uses a chunkier grain. Note that some of the small connecting lakes are now invisible, having fallen below the threshold of the grain. In C, the pixels that constitute the grain are one kilometer, and so most of the detail from the other two images has been lost in the coarse grain. *From B. J. Benson and M. D. MacKenzie, "Effects of Sensor Spatial Resolution on Landscape Structure Parameters,"* Landscape Ecology *10 (1995): 113–20. Reproduced with permission.*

An example that has both spatial and temporal grain and extent is the set of images from space seen on the nightly news weather report. Here a single image is taken at a certain time, but several images are then run together to show dynamical change throughout the past day. The temporal grain is the time between images, as they are shown in sequence, and the extent is the length of time covered by the entire sequence. Swirls of low pressure regions appear jerky because the grain is fairly coarse. If finer-grain measurements were integrated, then the patterns of movement would appear smooth.

Temporal scale refers to *frequency of behavior*. Behavioral frequency is the amount of time it takes for a cycle to be completed and start again. Fast behavior is high-frequency, and slow behavior is low-frequency. Particular types of things behave with particular *characteristic frequencies*. Typically, physically smaller things have a short return time, or high characteristic frequency, while large things have a long return time, or low

In D and E, images from the NOAA weather satellite were pulled from the
University of Wisconsin ethernet at the time we were writing this figure legend.
D: April 5, 1995, at 23.15 hours. E: April 6, 1995, at 15.45 hours. The temporal
grain is thus sixteen hours, and the spatial grain is uncertain but is at least five
miles, whereas the spatial extent is North America. The large cloudless area is
over the Florida panhandle. The storm system sweeping diagonally from lower
left to the upper center of the images made it rain in Wisconsin quite hard that
day. The NOAA images came through the Environmental Remote Sensing
Center and Department of Meteorology, University of Wisconsin.
*Courtesy of B. Sherman, who processed the images to make clouds white and overlaid
the state outlines.*

characteristic frequency. For example, whale cells divide faster than whales reproduce. Whales are spatially large, low-frequency entities that have a longer return time than whale cells. Of course, a different emphasis in what is considered important might reverse relative rates. For example, the constancy of genes inside the whale's cells that is passed on to offspring may be recognized as existing much longer than a whale's life span. The animal acts as merely a temporary home for the continuing genetic signal. Fast and slow are relative terms, which need to be interpreted from an observer's point of view, and interest in a phenomenon.

The Tradeoff Between Grain and Extent

We have just seen that all observation is subject to limits inherent in the sampling method, and occurs within the context of an undifferentiated but nonetheless potentially influential background. No observation will capture all there is to be seen, and that which is too big or too small remains unsampled. A complementary relationship exists between grain and extent. Often, difficult measurement is due to competition between measurement grain and extent. High-frequency behavior demands frequent measurement. In order to observe short cycles one must look often, but there is an associated cost of accumulating measurements. If differences between measurements count, then there is only so long one can make fine-grain measurements before, buried in data, one loses track of what is happening. Thus measurements that are close in space or time usually limit the width of the span captured by the extent. Accordingly, a fine-grain measurement regime is most suitable for a restricted, or narrow, extent. As the level of detail associated with the grain increases, the extent accordingly decreases. Fine grain narrows scope.

Conversely, as the extent of an observation protocol increases, the level of detail captured in the grain must decrease. If it takes a very long time to capture a low-frequency phenomenon, then frequent measurements are a waste of resources. Hourly measurements of temperature are inappropriate if one is searching for an ice age. Such closely spaced but distinct measurements are ineffective. Low-frequency events, which rarely occur, impose a coarser grain on observation. Accordingly, a coarse-grain measurement regime is most suitable for a broad extent. Wide scope sacrifices details.

FIGURE 3.8 "Bubble chambers are the devices used by physicists for detecting subatomic particles." *Drawn by Kandis Elliot*

FIGURE 3.9 "Nevertheless, 'chairness' hinges on criteria embedded in cognitive models." *Photograph by Paula Lerner*

Competition for attention between detail and scope forces decisions as to what can be entertained in data collection. On the one hand is the adage "can't see the forest for the trees." On the other hand, to reverse the adage, "By focusing on the forest you lose track of the trees." Computers relax the tension between grain and extent by allowing more details to be remembered than would otherwise be possible. Even so, we often press the limits of computation such that the tradeoff between grain and extent returns to force hard decisions. Remarkably quickly, details are encountered in such numbers as to exceed computational power.

The challenge of very difficult measurements can often be expressed as a conflict between simultaneous demands for a fine grain and broad extent. For example, some fundamental particles exist in theory, but their effects have been observed only in bubble chambers after a decade of intensive observation. Bubble chambers are the devices used by physicists for detecting subatomic particles (figure 3.8). The reason these particles were so difficult to see is that each image needs careful scrutiny for the effect to be seen, but the critical event happens only every once in a long while. A wide extent and fine grain are required to observe the phenomenon, and this combination is very difficult to achieve.

Capturing the spatial and temporal/frequency properties of external events requires measurement that is simultaneously wide enough and narrow enough for the observed to fit within the scalar limits prescribed by the measurement protocol. Grain and extent set scale, and limit which entities and cycles may be observed. Not seeing something could be due to its not being there, or it could be due to inappropriate measurement because of ineptly chosen grain and extent. Effective measurement brackets the spatial and temporal characteristics of that which is sought.

Internalities and Externalities

Models versus Dynamics

We now turn to the issue of relations between structure and scale. We presume that the changes of state that arise in observed structures reflect an externally derived dynamics. Dynamics are *rate-dependent*. Structure, on the other hand, arising as it does from the observer's criteria for things, is *rate-independent*. Once again we return to the mun-

FIGURE 3.10 Spatial size and characteristic frequency have nothing to do with being of the type called "organism."

dane to illustrate rate dependence and rate independence. Consider once more a chair. The chair's behavior, such as falling over, occurs in space and time. Nevertheless, "chairness" hinges on criteria embedded in cognitive models.

   Criteria are internal to cognitive models and exist independently of the spatial and temporal properties of the observed. Something being recognized as a chair is a call of judgment on the part of the observer. In looking for a chair, one is looking for a rate-independent structure whose scale has not yet been prescribed. On the other hand, what one in fact sees is a particular scaled entity, perhaps something one is prepared to admit as a chair, whose dynamics are rate-dependent. Before observation, the parts of the structure may have a generalized scale-relative relationship, like "bigger than," although one must wait until after an example has been found before the precise scalar dimensions of the parts are identified. Across these differences, before and after the act of observing, scale and structure interface in observation. Forging a link between scale and structure, without confusing the contributions of the observer and the observed, is at the heart of hierarchy theory.

We have been imagining the world as in constant dynamical flux. In the face of pure flux, what makes definitions and criteria powerful is that they isolate repeating patterns. When a label is attached to a pattern, the effect is to make the dynamic become static. In other words, part of the flux is frozen to give rate-independent structure. While our experiences of the world are linked to change over time (dynamics), our definitions, which correlate with patterns of repeating experiences, are static and available to serve as elements in cognitive models. To return to our fishing metaphor, we cast our line into the flowing stream, and any structural entities caught, such as fish, get put into our bucket for later assimilation.

In chapter 2 we described the observer's criteria for focal entities as an early stage of observation. A priori criteria, which make an entity stand out from its background, reside entirely in the observer's cognitive models. Consider, for example, the criteria for an organism. The criterion might be 1) cells of a biological structure all have the same genetics because they derive from a single sexual union; 2) the structure is all in one piece; 3) as a whole, the entity is separate from others of its type; 4) its cells are genetically different from those of others of its type; and 5) there is a physiological coherence to the thing under observation. According to these criteria both mites and whales are organisms. These five criteria bring into the foreground a set of relationships that isolate certain figures from an undifferentiated, dynamical background. Both a mite and a whale meet the criteria for organisms despite very different physical sizes, and rates that give very different characteristic frequencies. Spatial size and characteristic frequency have nothing to do with being of the type called "organism."

Knowing that something is called an "organism" does not give much indication of its spatial or temporal characters. It does suggest that the

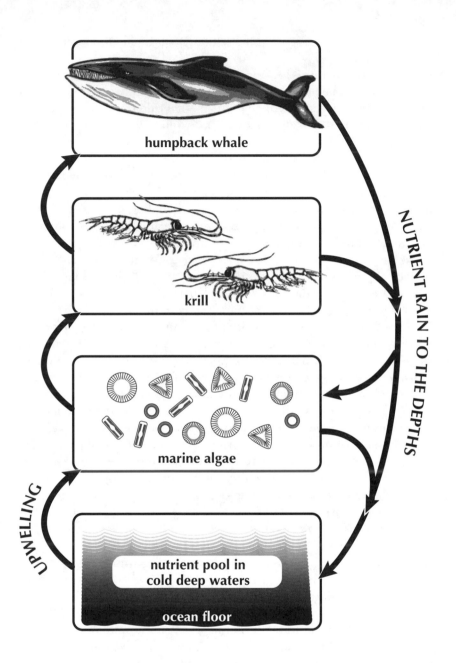

humpback whale

krill

marine algae

nutrient pool in
cold deep waters

ocean floor

NUTRIENT RAIN TO THE DEPTHS

UPWELLING

entity in question is at least as large as a virus but no bigger than a red-wood tree, and no older than ten thousand years. If it is an organism, it is a safe bet that it is larger than a protein molecule and smaller than a mountain. There is a lot of leeway as to the actual scalar properties within the general scalar limits of being an organism.

There is also enormous leeway as to how one characterizes a given material entity. Just because something would fit the criterion for being an organism, there is no requirement that the observer makes such an assignment. Consider, once again, a whale. The material system that cor-responds to a whale need not be recognized as just an organism. If we instead pay attention to the fact that the whale consumes and breaks down krill, then it is a secondary consumption compartment in an ecosystem. If we recognize that much of the material consumed is fairly quickly defecated and urinated out of the whale, then it becomes just a link in a nutrient cycle. From such a perspective, the whale need not be seen as significantly different from the pool of nutrient-rich water at the bottom of the ocean, which is also part of the nutrient cycle in the oceanic ecosystem (figure 3.11).

Considering a whale as a secondary consumption compartment, or as a link in a nutrient cycle are both independent of spatial and temporal considerations. Units ranging from fish in aquarium to biota in aquar-ium, or from whale as organism to whale as secondary consumption compartment, are the product of the observer's structural decisions of "thing" versus "not-thing." All these criteria are equally valid because they are erected a priori, that is, before experience. As such, criteria are independent of temporal and spatial considerations. They are formal, and revising or changing units is a matter of weighing the costs and ben-efits of adapting this as opposed to that criterion given the investigation at hand.

While the criterion for organism is independent of space and time, many features of organisms do require taking into consideration scale-dependent properties. For example, characteristic frequency, spatial size, territorial range, and life expectancy of a particular organism are all

FIGURE 3.11 A whale can be seen as a discrete entity that eats other entities in a food chain. However, whales hold nutrients a long time, so from a different perspective one could unite the top and bottom compartments here—the whales and the deep-ocean nutrient pool—and call the new amalgam the "nutrient store."

FIGURE 3.12 "Looking for some mythical animal, like a dragon." *Photograph by Paula Lerner*

scale-dependent properties. They are all post hoc, that is, discovered after the fact of observation. As the contribution of the observer, criteria can be changed at will. However, one cannot change the spatial and temporal properties of the observed. While equally important, post hoc properties are observed rather than postulated. They are the contribution of the observed, and beyond the observer's control.

Definitional versus Empirical Entities

The immediately preceding discussion leads naturally to the recognition of two sorts of entities, one type definitional and the other empirical. *Definitional entities* are postulated before a measurement is made. They are what you are looking for when you begin. Definitions can exist whether or not an example that matches the definition is there to be found. Knowing what you are looking for does not have to mean that you will find it. I may, for example, set out looking for some mythical animal, like a dragon or unicorn. I may know what I mean by a unicorn, and I may search for unicorns, but that does not mean that I will find one. Definitional entities are important because you cannot see anything systematically until you pose a question, and posing a question invokes assumptions about what could be seen. Definitional entities are observer-generated, and they fix the level of observation at the outset.

Knowing what I am looking for does not mean that I will find it, for I may find something else. Once a measurement is made, a new class of entities arrives on the scene. What is actually seen is a second class of post hoc, *empirical entities.* Perhaps I look for unicorns, and instead discover horses. The net of observation is cast, and structured experiences are its yield. Horses are similar to unicorns in size and behavior, and so they fit within the net intended for unicorns. Perhaps I did not find what I went looking for, but found something equally interesting instead (figure 3.13).

While definitions are scale and rate-independent, that does not mean that experience is excluded in the process of revising definitions. After discovering a horse instead of a unicorn, I am free to modify my definition; it changes from four-legged, single- not cloven-hoofed mammal with one horn to a similar four-legged mammal without a horn. I am also free to change my question, now that I am interested in learning about horses instead of unicorns. My net of observation yielded something I

FIGURE 3.13   "I look for unicorns, and instead discover horses."
Drawn by Kandis Elliot

had not anticipated at the outset, and this discovery provoked me to create a new category, bounded by a new criterion.

Modifying a definition so that it better fits experience should not be
interpreted as an internal definition becoming external. Definitions are
always completely internal. Modifying criteria may be guided by pragmatic considerations, such as making the definition more parsimonious
or relevant to what is actually observed. Becoming more parsimonious or
relevant is not, however, the same as becoming more true. Definitions
are not true or false, they are only more or less useful. Because definitions are embedded in internal cognitive models, their worth hinges on
pragmatic rather than ontological considerations.

In summary, definitional entities are postulated before observation and guide the process. They are completely observer-dependent, and scale- and rate-independent. Armed with entities postulated in a question and a measurement protocol of fixed scale, empirical entities are discovered in the process of observing. Which empirical entities are discovered may not, however, have been anticipated at the outset. They are experience-derived, and their temporal and spatial characteristics, which necessarily fit within the limits of the measurement scale, have to matter. Having stumbled upon an empirical entity, the observer is then free to modify definitional criteria for future observation. However, the new criteria are not any less observer-dependent. Criteria are always observer-derived, even though they are modifiable. Empirical entities are experience-derived, and their rate- and scale-dependent properties are beyond the observer's control.

Reification of Empirical Entities and Realism

Treating definitional entities as empirical entities, or empirical entities as definitional entities, suppresses the fact that experience is always mediated by concepts and measurement scale. The observer's perspective and measurement decisions are often implicit and easily forgotten. Once perspective and measurement decisions are taken for granted and then forgotten, one is easily seduced by the mistaken belief that definitions are discovered in nature rather than actively erected by an observer. We should highlight two errors that arise in contrast to our position. The two mistaken positions are *reification* and *realism*, and our contrasting position is *constructivism*. In reification, the seduction takes the form of considering that something is real because one has asserted its existence. Under reification, one believes something is real because humans have decided upon it. Of course, that is not how one who reifies would state it, but there are pressing arguments to indicate that such is what happens in reification. An example of reification might lead a biologist to assert that "all this ecosystem nonsense is just an abstraction; I study organisms, and at least they are real." Botanists know better; but zoologists are inclined to assert that species are real, when other levels like genus or family are supposed to be mere human abstractions. Reification is a primitive philosophical position, that of naive realism, and many reifiers do not notice that they are doing it.

Realism takes a form that is in a sense the reverse of reification. While the order comes from humans in a reification, realism comes from the material system. It is nevertheless still at odds with the position we recommend. Under realism, there are real categories, and it is these categories that give rise to human categories. Plato was a realist, so the position has some credentials. The world is full of things *as things*, and humans supposedly identify these real things after experiencing them.

Our position is antirealist, but it does not assert solipsism, a philosophy that states that all is a matter of human construction. Under solipsism, there is no world out there, for everything is a human construction. For the solipsist, since we could never have access to the real world and would not know it if it existed and we found it, there is no point in being constrained by a presumed material system. Rude excesses of postmodernism come from this view, where almost all science, history, and literary interpretation is an elitist conspiracy, and nothing can ever have significance, particularly dead white men like Napoleon or Shakespeare. Our position is that there probably is a world where there is existence, but that things do not exist as things out there. We never have access to the world as such but learn from interaction with it. Knowledge is constructed from our interaction with observables. Under constructivism, the things that emerge from the interaction are a product of the interaction, not some prior existence as things. A different interaction would produce a different set of categories. Constructivists would say that Newton's gravitational laws are not necessary, although they are a very good model; another intellectual framework deriving from a different constructivist interaction would lead to a different codification. Gravity, as such, is not out there waiting to be discovered, although there is some indication that Newton did a good job, out of an undefined number of other good jobs he might have done.

External, undifferentiated dynamics provides the fodder for experience. In order for categories to exist *as categories*, and things to exist *as things*, there must be an observer building interpretive models. Model building is not, however, a matter of waiting for one's mind to resonate to the world's "natural" categories. Definitions and criteria are not discovered in nature. Instead, they are the product of an active observer's search, given a particular intent and perspective.

Criteria give structure to and interpret experience and are modified in order to facilitate assimilating new experiences into existing cognitive models (the observer's web of belief). In the absence of cognitive models, external dynamics possess no structure or meaning. Accordingly, categories and their members do not exist externally. Put more strongly, science is not about finding out the truth. Science is about socially acceptable perception. Instead of arguing about the truth or falsity of definitions, or about what is "really" out there, arguments are more profitably couched in terms of whether a definition is useful or misguided, and whether or not one's categories map neatly onto repeating patterns of external dynamics.

Conclusion

In this chapter we looked at the relations between measurement scale and definitional and empirical entities. In an observation, the contribution of the observer is criteria for things that might be seen, and choice of measurement grain and extent to bracket the scale of observation. Criteria are internal and rate-independent. Observed, empirical entities are experienced in space and time and their behavior exhibits rate dependence. Empirical entities fit within the observer's measurement scale. Their temporal and spatial features matter. With these distinctions in hand, we are now in a position to look at how spatial and temporal characteristics of empirical entities provide one metric for ordering levels of observation.

4

Ordering Hierarchical Levels

In the last chapter we identified the distinction between definitional entities that are erected a priori and observed entities that are found empirically. The entities discussed so far were considered, to a large extent, in isolation. This does not invite an easy discussion of levels, because individual levels are occupied by more than one entity, and multiple levels necessarily must consider at least one entity for each level. A discourse focused on levels invites consideration of multiple entities, be they members of the same or different levels. What makes levels interesting is the relationships between them. Having clarified different classes of entity, we are now in a position to look at the relationships between levels that are characterized by their respective entities. Levels are characterized by the entities that reside in them. Accordingly, levels fall helpfully into two categories defined by the two types of entities distinguished in the last chapter: those entities and levels that derive from definitions, and those entities and levels that result from empirical observation.

Levels of Organization versus Levels of Observation

Introductory biology is often arranged from cells to large ecological systems and the biosphere or from large ecosystems to microscopic cells (figure 4.1). Sometimes this works well enough, but it is probably not worth the confusion it invites. While individual organisms in a population are smaller than the population to which they belong, many organ-

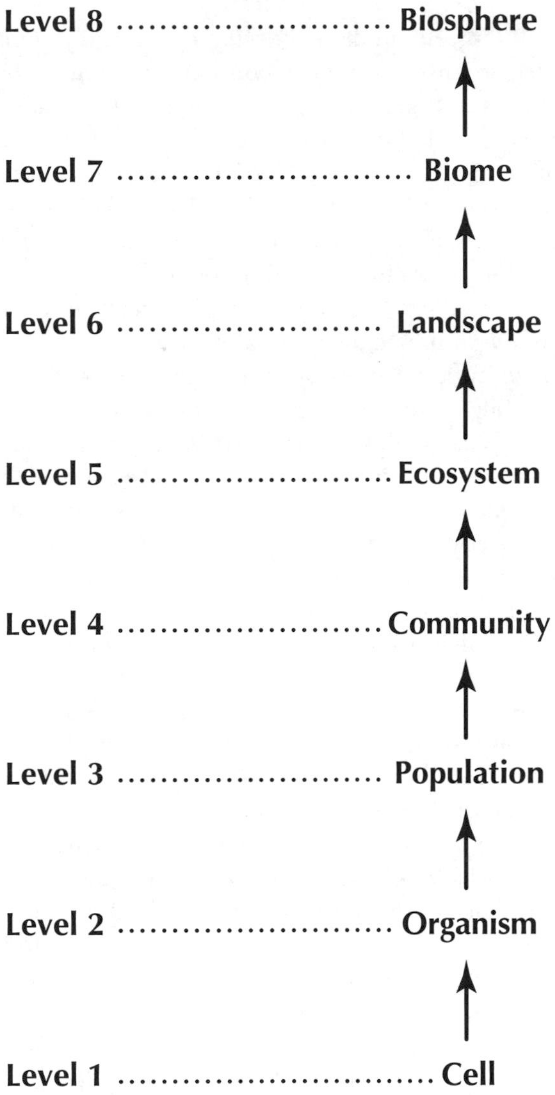

Level 8 ·························· Biosphere

↑

Level 7 ··························· Biome

↑

Level 6 ························ Landscape

↑

Level 5 ························· Ecosystem

↑

Level 4 ························ Community

↑

Level 3 ························ Population

↑

Level 2 ·························· Organism

↑

Level 1 ···························· Cell

FIGURE 4.1 The conventional levels of organization as they commonly appear, ordered to outline an introductory textbook. It is not so much wrong as it lacks generality. It captures very little of the material flows in biological material.

isms are much larger than many populations. For a population of parasites, the host organism is a large contextual resource base. Thus the hierarchy of life is not so much wrong, as it applies to a very small subset of the relationships that occur in material flows around the biosphere. The hierarchy of life model is an ordering based on formal inclusion relations. For example, organs contain cells, organisms contain organs, species contain organisms, communities contain species, and so on. It is an ideal ordering according to definitional criteria, not a description of a particular material system that can be empirically uncovered.

It is easy to overestimate the generality of the scalar ordering found in interlocking definitions. In the conventional biological hierarchy, scalar order follows only in the case of particular examples. Each general category applies to a range of very differently scaled entities. For example, a leaf is a landscape for mites that feed on leaves, and organisms range from microscopic to leviathan. Therefore the general categories in the biological hierarchy cannot be ordered by spatial extent in a general way, because they depend on the size of the particular exemplars employed as to which is larger. For example the landscape or organism in question will determine which is the larger, and so determine which level, landscape or organism, is the context of the other. In some cases, as in the host parasite example above, the host organism is a landscape.

To take another example, let us return to whales. It makes intuitive sense that whale cells occupy a lower level than do whole whales, but not all cells are smaller than all organisms. Disease in whales could involve large, entire populations of bacteria that are nevertheless much smaller than the whale cells on which they depend. The host cell belonging to a whale becomes the context of the entire bacterial population. It is possible to observe populations of organisms like barnacles and discover that entire barnacle populations are smaller than an organ of a single organism, like the whale skins on which they live.

Building models based on definitions is a purely logical exercise. While hierarchy theory can be useful in clarifying levels related by definition, it can do more by offering a method for ordering empirical entities into levels of observation. Different levels in an empirical hierarchy are populated by entities that differ in size and frequency characteristics. These features are linked to the measurement scale. The great advantage of scaled empirical levels, over those ordered by definitions, is that the order of scalar levels corresponds richly to many other orders and relationships that the observer did not necessarily have in mind when erecting definitions. Thus empirical hierarchies can be powerful exploratory devices. Definitional hierarchies, by contrast, tend to be most useful either in getting the first observational framework started when one is still profoundly ignorant or when tidying up observations post hoc.

In this chapter we offer an alternative to the traditional hierarchy of life. We erect one based on spatial and temporal characteristics of observed, empirical entities. In the contrasting schemes, definitional entities are ordered into *levels of organization*, while empirical entities are ordered into *levels of observation*. For example, dog-canine-carnivore is a logical hierarchy based on formal, inclusion relations among the sets at each level. Each category is defined by criteria, and each level represents a different logical type. The categories within this hierarchy are internal, cognitive constructs that do not depend on experience or observation. An equally valid hierarchy is unicorn-animal-organic, even though no one has actually seen a unicorn, because there are none to be seen. The unicorn hierarchy is valid because it rests on observer-generated criteria that are erected before experience.

Ordering Levels of Organization and Observation

Hierarchical models organize either levels of organization or levels of observation. While a definitional hierarchy can be consonant with an empirical hierarchy, power is severely limited, and analyses become muddled if definitional and scalar levels are confused one with another. In a hierarchy of definitional levels of organization, levels are ordered according to entailment and inclusion relations implicit in the criteria themselves. In a hierarchy of empirical levels of observation, levels are ordered according to the spatial and frequency characteristics of the entities that occupy each level.

Distinguishing between definitional and empirical hierarchies is important. Data may or may not be relevant, but one thing is certain: data alone will never be able to differentiate between which of two competing definitions is superior. The only way to choose between alternative definitions is by evaluating the role they play in models. Changing definitions is a matter of pragmatic concerns such as: which model is the most parsimonious, or which is the most inclusive of experience? The debate may point to an insufficiency of relevant data, but the data by themselves neither build nor modify cognitive models. Even the use of statistical tests applied to data do not tell the scientist what to do; they

only give probabilities of the consequences of making this as opposed to that decision. The terminology of statistics gives the clue: tests of significance. Significance is a matter of judgment. Decisions as to the adequacy of definitions do not turn on data, they turn on judgment.

By contrast, resolving conflicts in empirical hierarchies involves more than revising definitions. It may involve incorporating new levels into the scheme to accommodate empirical differences, or it may be a matter of acknowledging the effects of different measurement scales. If arguments cannot be resolved satisfactorily given present knowledge, then there are measurements that can be made to indicate which set of proposed relationships gives a superior description. Conflict in empirical hierarchies can often be settled by data, as the name "empirical hierarchy" would suggest. A muddling of empirical and definitional hierarchies leads to arguments as to the correctness of apparently conflicting data, when in fact data cannot resolve the conflict.

Science focuses on settling questions by careful thought combined with critical measurements. Sorting out the difference between debates where collecting more data will help and those where data are irrelevant is at the heart of many debates regarding the nature of complex systems.

Ordering Levels of Observation: An Example

An example will help illustrate how levels of observation are populated by empirical entities scaled according to different spatial and temporal frequency characteristics. One way to understand complexity is to simulate something complex, and then look at how the simulation works. The difficulty here is, first, to know whether the system being simulated is indeed complex, and second, to be sure that the simulation captures the essential relations within the material system. One system that fits both of these criteria is human conversation. We all know firsthand that humans are complex, and we know when a simulation has got it right, because a computer holding a conversation makes one feel most uneasy.

ELIZA is a computer program that behaves like a Rogerian therapist. It is sufficiently realistic that it is even used as a therapeutic tool; some clients apparently open up to a machine more readily than they do to a human doctor. A Rogerian is not from the Freudian "Aha!" school of

FIGURE 4.2 "One system that fits both these criteria is human conversation."
Photograph by Paula Lerner

therapy that digs down into your subconscious to find a crucial trauma. Rather, the Rogerian therapist tries to turn emotional problems back on the client. Rogerians appear very agreeable and passive as they ask non-committal questions about family and other relationships.

The following computer output has been published before in several places, including T. F. H. Allen's and T. B. Starr's *Hierarchy: Perspectives for Ecological Complexity*. The program really does perform as in this disconcerting example: we know, we tried it. We typed in the same words as those used by the client in the conversation, and the responses were identical to those in the published account, except for the last line. We stalled for a while making ad lib comments in the same vein, and eventually it even gave the last line verbatim. While we were waiting, the program made several attempts to double back and make us look at the big picture, with comments like EARLIER YOU MENTIONED SOMETHING ABOUT YOUR FATHER. The last line is a question that turns back to the top of the conversations and appears to be an implementation of a standard strategy of the program. Here is the conversation with ELIZA.

Men are all alike.
IN WHAT WAY
Well, they are always bugging us about something.
CAN YOU THINK OF A SPECIFIC EXAMPLE
Well, my boyfriend made me come here.
YOUR BOYFRIEND MADE YOU COME HERE
He says I am depressed.
I AM SORRY TO HEAR THAT YOU ARE DEPRESSED
Yes, I need some help, that much seems certain.
WHAT WOULD IT MEAN TO YOU IF YOU COULD GET SOME HELP
maybe I could learn to get along with my mother.
TELL ME MORE ABOUT YOUR FAMILY
My mother takes care of me.
WHO ELSE IN YOUR FAMILY TAKES CARE OF YOU
My father.
YOUR FATHER
You are like my father in some ways.
WHAT RESEMBLANCE DO YOU SEE
Well, you are not very aggressive, but I don't think you want me to notice that.
WHY DO YOU THINK I AM NOT VERY AGGRESSIVE

Because you don't argue with me.
WHY DO YOU THINK I DON'T ARGUE WITH YOU
Because you are scared of me.
DOES IT PLEASE YOU TO THINK I AM SCARED OF YOU
My father is scared of everyone.
WHAT ELSE DO YOU THINK OF WHEN YOU THINK OF YOUR FATHER
Bullies.
DOES THAT HAVE ANYTHING TO DO WITH THE FACT THAT YOUR
BOYFRIEND MADE YOU COME HERE

One almost wonders if the computer knows that women marry their father figures and men marry their mothers. Can it really know about macho American males? The answer must be no, but then one still has to explain how it appears to reason as it does.

This remarkable behavior on the part of a machine can be seen as employing several levels of observation. Remember that low levels are characterized by small-scale entities, that is, ones that are either spatially small or behave quickly. An entity at a low level of observation completes its cycle of behavior in a short time period. It displays high-frequency behavior. Conversely, high levels of observation are populated by large, ponderous, entities displaying low-frequency behavior. In the simulation, the computer responds with different periodicities. Some responses are rapid-fire, whereas others can only be understood as part of a longer sequence of lines. The different lengths of time taken to complete a cycle in the conversation represent different levels of observation. They require different numbers of lines of text to see the patterns of repetition.

The briefest responses are the lowest level. They are only an echo of one of the client's statements. For example, the comment, "My father," is met with YOUR FATHER. This sort of one-line utterance followed by a one-line response occurs several times in the text. It serves the purpose of reassuring the client that the computer really is listening. We all do this in conversation with nods of the head or a low-key "uh-huh" noise. This is an important but low level that conveys little information, and does not convince one that the simulation is adequate or complex. However, with these tailor-made quick-fire responses, the simulation cannot be explained away as some generic, canned set of responses that are the same for all clients.

At the next level there are key words, usually emanating from the client. These specific words are met with their generic counterparts. For example, the word "mother" is rejoined by its general case FAMILY.

Because "family" is a more general word, it can be used more than once without the computer saying something stupid. It takes a full five lines for the FAMILY response to work itself through the conversation. This middle-level behavior is the main substance of the conversation. As a rule, middle levels characterize any system most richly.

Finally, there is an even higher level of behavior. It is this level that gives the computer the appearance of complex behavior and makes it so uncannily human. The highest level makes it appear that the computer properly understands the significance of the whole conversation. The computer has somehow identified the context for the whole piece, and it shows this with the final rejoinder, DOES THAT HAVE ANYTHING TO DO WITH THE FACT THAT YOUR BOYFRIEND MADE YOU COME HERE. Of course it has everything to do with it! The poor client was apparently driven crazy by the power structure in her upbringing, and now it spills over into her relationships outside the family.

The remarkable, counterintuitive insight to be had here is that the computer does not need to know any of the details of the conversation to be able to make a connection between fine-grain responses and the larger context. All it has to do is identify the context, and the problem is solved. The solution is easy, and does not require any attention to the specifics of what is said. If someone has an emotional problem, then it is either a person or it is manifested by a person, and the most important person will be mentioned first, preceded by the word "my."

For the first few lines, the simulation is waiting for that person to be identified. Notice how it asks a pair of innocent questions to get the information. Once it has found out that the boyfriend is central, then it does nothing for a while but count the lines of the exchange. Finally the program makes its move and springs the BOYFRIEND question. Note how this upper level is remarkably independent of the lower-level details. Most of the force of the connection between her attitude toward bullies and being bullied is in fact supplied by you, the observer. As an experiment, try putting the BOYFRIEND question somewhere else in the passage and you will find that most of the time it still makes some large and telling point. That point may be different in detail from the one made when it follows "Bullies," but the same general message comes through.

If you did not know that one of the parties was a computer, you would probably interpret this conversation at face value. It would appear that a therapist is keeping track of an interaction at several levels simultaneously, and is integrating information across those levels. All the while the therapist makes polite "I am listening" noises, and asks questions about the family. Throughout the encounter the psychologist is interpreting responses in the context of an impression gained at the outset, namely that the client has problems with being dominated and manipulated. The complexity comes from fine-grain detail, such as the one word "bullies" being directly related to the larger context of a recursion pertaining to the whole conversation. In fact, since the computer cannot make that link on the basis of cultural understanding, the tightness of the relationship between the boyfriend and bullying is illusory.

How does the computer keep track of all the detailed implications of each word in the layers of meaning? The answer is that it does not! But then again, neither does a real Rogerian therapist. The human professional would not plan to steer the encounter toward the particular final rejoinder any more than would the computer. A real Rogerian might well simply make a remark about the boyfriend and work out the implications only afterward.

We define complexity as an apparent connection between fine-grain structure and behavior and large-scale structure and behavior. Large-scale structure corresponds to events widely separated in space or time, which are nonetheless coherent. In the present example, the larger scale is the context of dominance and submission in interpersonal relations, and the fine-grain detail pertains to this particular boyfriend as a member of the class "bullies." One is commonly amazed that a complex system, whatever it is, can keep track of all that accumulation of detail which occurs while the large-scale aspects of the system are turning over. As can be seen from the present example, the key is precisely that the larger context does not in fact keep track of accumulated details. Yes, complexity requires that a link be made between details and context, but the details reside at a low level and do not play a simple, identifiable role in the larger, causal connections at the level of the context.

Much of the integrity of complex systems is with hindsight of the observer. Although there is some relationship and integrity between levels, separate levels function in a remarkably autonomous fashion. Complexity is not just intricate, it is usually also loose.

Simplifying Multileveled Systems

The ELIZA computer program is insightful and shows that complexity can be made tractable by reducing it to an interaction between empirical

Empiricism

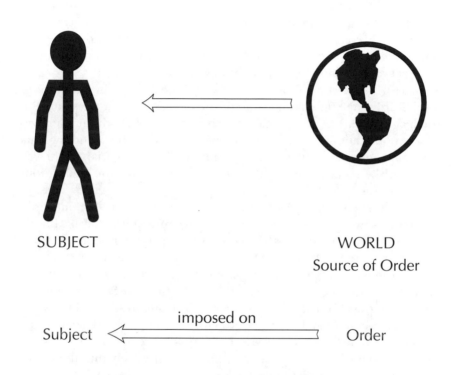

SUBJECT

WORLD
Source of Order

imposed on

Subject ⟵========= Order

Development = accumulating experience

FIGURE 4.3 The nature/nurture debate turns on taking either an empiricist or nativist point of view. The constructivist position that we take has already been presented in figure 1.9. The constructivist view indicates that the whole debate lacks substance at any level.

Nativism

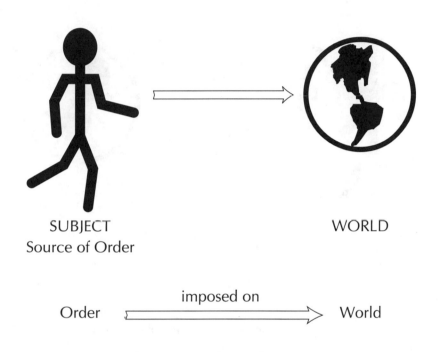

SUBJECT   WORLD
Source of Order

Order $\xrightarrow{\text{imposed on}}$ World

Development = physical maturation

levels of observation. Notice that this type of analysis is a very different solution from reducing a multileveled system to a small set of variables, over which tight control is exercised. The goal is not to eliminate multiple levels, but rather to identify their boundaries and infer causal relations between them. Addressing complexity one level at a time breaks the phenomenon into components that are still manageable in terms of memory and information processing. After that analysis, one has a grasp

Constructivism

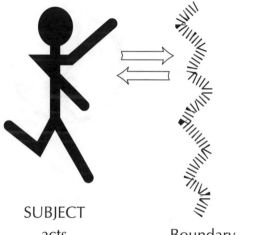

SUBJECT   WORLD
acts   Boundary   reacts

Interaction = source of Order

Development = progressive change in subject as
              function of interaction with the world

of the levels individually, and can begin to piece them back together into
a whole. We should emphasize that putting them back together is not a
matter of simple summation, because different levels involve differently
scaled subsystems. In the ELIZA simulation the different levels are not
the simple sum of lower levels. A mere collection of all the "my father"/

FIGURE 4.4 "An epiphyte is a smaller plant that sits on top of another plant, like orchids or pineapple relatives on trees in a tropical forest."
Drawing by Kandis Elliot

YOUR FATHER low-level exchanges would miss the significance of the entire conversation.

A further complication is that there is no absolute basis for a level's position as high or low in a hierarchy. Level position is always relative to the other levels with which it is compared. Change the basis for comparison, and entities that belonged to high levels might now belong to lower levels. Scale is a matter of measurement, and it should be clear by now

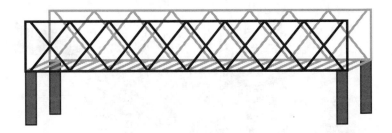

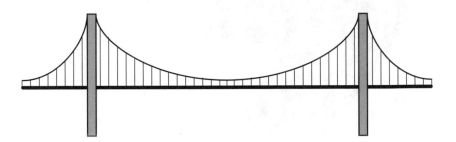

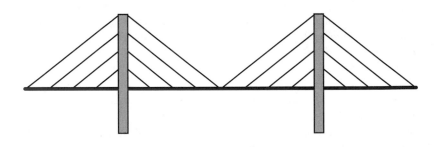

that measurement is heavily burdened with observer decisions. Indeed, there is no general prescription for the choices of levels in taking apart complex behavior.

We have defined levels of observation as differing based on the characteristics of empirical entities that populate them. The general rule is: higher levels of organization are populated by entities that are spatially large, or whose characteristic behavior is low-frequency; and lower levels of observation are populated by spatially small or high-frequency entities. This general rule allows empirical levels of observation to be ordered relative to one another.

Empirical Hierarchy of Life

At the beginning of this chapter we complained that the conventional hierarchy of life is often mistaken for an empirical hierarchy when it is in fact a definitional hierarchy. Having drawn out the tension between definitional and empirical hierarchies, we are now in a position to offer a remedy for the conventional hierarchy of life. Our proposition is an alternative to the levels of organization in the hierarchy of life. Instead of ordering levels according to definitions, one could order levels according to spatial or frequency characteristics of the observed. We recommend turning more often to levels of observation.

In their book *Toward a Unified Ecology*, T. F. H. Allen and T. W. Hoekstra suggest a model for moving between levels in a biological hierarchy. If one separates the scale of an entity from its type, then it is possible to separate the definitional hierarchy of types of biological entities from an empirical hierarchy that is ordered on scalar properties of aspects of the

---

FIGURE 4.5 Bridges fall down regularly. By this we do not mean often, but rather that a new design is extended for about a quarter of a century, until it then fails because a particular example has been scaled outside the conditions for which it was designed. With steel as a medium, stone arches were replaced by steel box construction, cantilevered. The Tay Bridge disaster occurred when steel box construction got too large. Suspension bridges were one of the replacement designs. At present, the design on the bottom panel is preferred. One of these will probably be designed outside the intended scale and fall down sometime in the next century.

FIGURE 4.6 The Tacoma Narrows Bridge turned like a ribbon in the breeze before the wind-storm took it down some hours later. *Courtesy of the* Tacoma News Tribune

material system. That approach is at odds with how most biology is studied, but it is the strategy we recommend here.

The conventional way to deal with mismatches between biological levels of organization and a scalar hierarchy is to erect a special vocabulary, one term for each type of exception. The conventional solution grabs the wrinkle in the hierarchy of levels of organization and stuffs it into a special-case definition. Thus reversals of scale in the hierarchy are encompassed in relationships forced by definition. This strategy of naming one's way out of an inconsistency works well enough for many problems, and is indeed a standard procedure for the development of many types of human cognition. Jean Piaget points out that children learning to deal with the world, as well as scientists seen through the window of history of science, both use typology for advancing their respective understandings (figure 4.3). In both infancy and science the strategy is to accumulate exceptions to whatever rules one is using, identify patterns therein, and create a new type to deal with this new class of happening. The new type is a metamodel that encompasses the old rule and its exceptions. The new model amounts to a set of definitions that fix relationships between levels of organization.

An example of this linguistic approach to biology is employed to deal with the reversal in the relative size of populations and organisms that occurs when large organisms meet large numbers of very small organisms. The standard solution is to invoke a concept like parasitism, a relationship between host and parasite that defines away the inconsistency with the conventional hierarchy that puts organism (host) above populations (parasites). Note that the scale of the host and the parasite is not stated, so there is room for ambiguity, but parasites are usually small relative to their host. Another example is the term *epiphyte*, which pertains to a similar problem of relative size, but not involving transfer of resources. An epiphyte is a smaller plant that sits on top of another plant, like orchids or pineapple relatives on trees in a tropical forest (figure 4.4).

Note that in both examples relative scale is part of the definition, but the precise scales for observation are not stated. Biology is full of terms that embody statements of relative scale without identifying any specific size or time span. These terms are formal models and can be generalized, as when organisms as different as bacteria and cuckoos are both called parasites. In brood parasitism, the host bird rears the young of the brood parasite who laid its egg in the host nest when it was unattended. As a result of these formal models, biology is full of rich analogies, where materially very different situations are nevertheless made comparable. However, these analogies and their terminology do not confront scaling effects in any particular material system. They clearly make limited use of scalar relationships between levels of observation and their associated material flows.

There is something to be said for borrowing an alternative strategy from engineers, one that addresses scale in biological systems directly. In engineering, as in biology, there are special terms that embody the relative scales of parts of the system—terms like tunnel and bridge. Nonetheless, in contrast to biologists, engineers spend most of their time dealing with explicitly scaled systems, like a bridge of a particular span. As systems get bigger, relative loading is calculated explicitly. When engineers fail, their mistakes are very tangible; the bridge falls down (figure 4.5). It might be the fault of the contractor not meeting the engineers' specifications. However, if it is a flaw in engineering, the engineer has succumbed to the intellectual strategy of biologists; the error arises from extending some formal model too far. Biologists are too often disconnected from material flows by a definitional framework that does not encourage scaled observation. When this happens to an engineer, the weaknesses of the approach become all too clear.

The collapses of bridges are not random, for they occur at regular intervals. The disaster happens when a design principle is taken so for granted that secondary effects, which could be safely ignored in the original use of a design, become suddenly and disastrously important when the design is applied elsewhere at a different scale. Having pressed suspension and other types of bridges to their limits, engineers now favor a new-style bridge that combines suspension and cantilevers. The span is supported by pairs of successively longer cables on either side of tall supporting towers, to give a bridge that looks like a row of Christmas trees. If history repeats itself, one of these bridges will collapse sometime early in the next millennium, when the design is pressed outside the envelope for which it was created. Some factor that was small and inconsequential will become unexpectedly important. In the case of suspension bridges, the Tacoma Narrows Bridge showed a harmonic wave in response to wind, with the roadway looking like a roller-coaster, behaving like a bull-whip in slow motion (figure 4.6).

For all their limitations, engineers are more reliable and predictive than biologists. While engineers can become overly focused on certain

FIGURE 4.7 "Napoleon was a small man but with far-reaching power."
Drawn by Kandis Elliot

formal models, they do spend much more time than biologists looking directly at scaling effects. Biology consists primarily of the accumulation of formal models collected into a massive vocabulary. Enrico Fermi once said that if he could remember all the names of the new particles proposed in modern physics, he would have become a botanist. Fair comment! The limitations of biology are characterized by its practitioners' predilection for more terminology. The weakness of biology is in its ready capitulation to definitional entities and levels of organization. Thus biology has a rich typology but a poor record of prediction in new situations, because it is slow to identify observed empirical entities and their associated levels of observation. The strength of engineering is its attention to direct scaling effects in systems of a particular size. Engineers use empirical entities and levels of observation as their default approach, and their predictions come true every time an airplane takes off or a building survives an earthquake.

In ecology there is a lot to be said for alternately employing levels of organization and levels of observation in an iterative fashion. The types of biological entity are definitional, but specific examples of each type occur at particular spatiotemporal levels of observation. There is no need to suppose that a given organism will be larger or smaller than a population of another species or a local ecosystem or landscape. Thus a move up- or downscale could lead to an entity from any definitional level of organization. On the other hand, it is also possible to keep the level of organization constant while taking pains to hold the type of definitional level constant. For example, a landscape ecologist might want to study landscapes, a definitional type of entity, while changing the spatiotemporal scale to encounter successively larger empirical entities at higher levels of observation. This is exactly what is happening when the ecologist measures the fractal dimension of landscapes, a measure of the complexity of a material landscape across a range of scales. Our recommendation is to keep levels of organization separate from levels of observation, but use them in tandem. In this way the observer has flexibility to propose any type of structure either as a context to give meaning from above or as an explanation from below, while simultaneously testing observables to erect predictive models.

Conclusion: Flexibility in Ordering Levels

By making a distinction between definitional and empirical entities, hier-

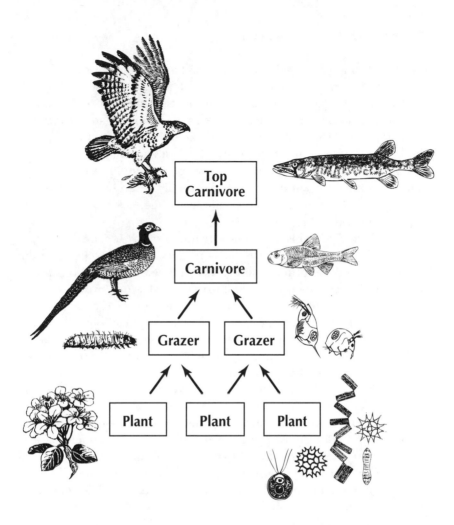

FIGURE 4.8 Food chains are normally drawn, as here, on the principle of the prey item being constrained by the animal that eats it. However, if the criterion is "depends on for food," then the wolf pack is seen as a satellite, dependent on the slow-moving, upper-level herd of ungulates. *Drawn by Kandis Elliot*

archy theory allows us the freedom to erect and change our definitions and measurement protocol so that they are efficient and useful in model building. A single object may have many facets, some small and fast, others large and slow. For example, Napoleon was a small man but with far-reaching power. Most often we would consider him a powerful person at a high level. However, this is true only if politics and military might are the ordering principles. A physically larger person could easily constrain such a small man. Clearly, any one empirical hierarchy captures only some aspects of nature, and which hierarchy one uses depends on the purposes at hand.

The ordering of levels in an empirical hierarchy depends on which aspect of the situation the observer considers significant. Change the significance, and the order of levels can change. A pair of levels populated by their respective classes of entities could switch orders without there being any contradiction, if the question used to construct the hierarchy changes. If the question is, "Which species depends on the other as a food supply?" then the deer are the upper-level context of the wolf. Conversely, if the question is, "Which species controls the number of the other through predation?" then the wolf is the upper contextual level for the deer (figure 4.8). The change from deer to wolf as the upper level comes from changing the point of view. That change identifies a different type of relationship between the two animals. The reordering from deer at the upper level to wolf at the upper level comes about because the new relationship gives a different order in terms of 1) frequency of behavior; and 2) context. These two criteria, sometimes accompanied by physical size, converge in the ordering of empirical levels of observation, such that different observers asking the same question can attain consensus on which ordering is most useful or appropriate.

Notice that scale, grain, extent, and levels of observation are all measurement issues. However, they operate in particular situations in the context of human decisions that derive from entity definitions and experience structured by prior scaling decisions. By being aware of the scale of one's investigation, the student of a complex system can shave off a level from the entirety of a phenomenon. Once each level has been analyzed, then the set of levels may be pieced back together according to their scale and observational criteria. The resulting hierarchy will be a coherent set of explanations that answer a particular question. The levels above give meaning, while the levels below explain

FIGURE 4.9 "The student of a complex system can shave off a level from the entirety of a phenomenon. Once each level has been analyzed, then the set of levels may be pieced back together." *Photograph by Paula Lerner*

by giving the origins of behavior. In the next chapter we look at other ordering principles for hierarchical structure. These principles often line up with scalar notions, like frequency of behavior and spatial or temporal extent, and so are a simple extension of the ideas that have been laid out in this chapter.

5

Changing the Context:
Constraint and Nestedness

Context and Constraint

So far we have argued that the scale of an observation is set by grain- and extent-measurement decisions. Whereas definitional entities are scale-independent, observed entities are scale-dependent. Finding observational entities at all, and certainly their character once one has found them, depends on measurement grain and extent. Levels of observation are populated by entities that have a particular spatial size and characteristic behavioral frequency. Levels of observation may be rank-ordered as higher or lower by comparing the spatial or temporal scale of empirical entities that occupy each level. In constructing a hierarchy of empirical levels, any given level fits between a higher and lower level, which are themselves populated by their own respective entities. Once hierarchical levels are ordered relative to one another, the next issue is to explore the functional interrelationships between levels. In this chapter we explore two types of relations between levels in an empirical hierarchy. One is control through constraint, and the second is nestedness or containment.

Control through Constraint

One source of system integrity is that higher levels are the context for lower levels. *Context* is defined as that which is constant when the system of interest exhibits behavior. Empirical entities have different aspects to them, some involving fast behavior, others involving slow behavior. Relationships between levels are defined by the particular aspects of the entities in question. Change the aspect that is considered critical, and

one changes the relationship. If a relationship is redefined so that new, relative behavior rates become important, it is possible for what was system to become context, and what was context to become system. This was addressed at the end of the last chapter as a reordering of levels.

By not behaving often, low-frequency entities are a relatively constant, upper-level context for higher-frequency entities. As an example of upper-level constancy, an organism such as a whale is the constant context of its cells. It is the low-frequency characteristics of large-scale entities that allows them to remain invariant while higher frequency entities are completing multiple cycles of behavior. Being low-frequency means not behaving often. As long as the background is relatively constant, it provides the context for the behavior of higher-frequency entities.

By being unresponsive, higher levels constrain and thereby impose general limits on the behavior of small-scale entities. Constraint is therefore achieved not by upper levels actively doing anything but rather by them doing nothing. For example, parents constrain a child's temper tantrum not by shouting back but rather by not reacting, and ignoring all the child's high-frequency thrashing and screaming. It may seem counterintuitive that imposing limits through constraint is passive, absence of behavior, rather than active manipulation. In the parlance of statistics, the upper level is a parameter for the variable behavior of lower levels. Parents are parameters for children's high-frequency developmental changes.

Physically small entities can populate high levels, but in that case the small entities must somehow offer a significant context for the spatially larger entities. This would apply particularly in control hierarchies, where a microcomputer represents the constant command structure over large engines. What defines a level as higher is that it serves as the context for lower-level entities. In this case, the smaller physical size of the control system is not relevant.

An Example of Ordering Levels

The relationships between behavioral frequency, context, and constraint can be illustrated by thinking about any hierarchic social structure, such

FIGURE 5.1 "Parents constrain a child's temper tantrum not by shouting back but rather by not reacting, and ignoring the child's high-frequency thrashing and screaming." *Photograph by Paula Lerner*

as the American legal system. It turns out that the founders of the United States, in using sound principles of government, adhered to the principles of hierarchy theory. For the purpose of this example we will divide the legal system into three levels: law enforcement by individual police officers, the state courts, and the federal courts.

The level of individual police officers, whose job may be described as monitoring the behavior of individual citizens, can be considered a low level of organization. Not only are individual humans spatially smaller than the region of jurisdiction of a court of law, but their behavior is also much higher-frequency. It is simply not feasible, even in totalitarian states, to monitor every behavior of every individual. The set of all behaviors of all individuals is too fine-grain to be captured by any centralized regime. Instead, the first step in the legal process is to differentiate between significant and insignificant behavior. Significant behaviors are phenomena, and in this case the phenomenon of interest is illegal behavior. Now the policeman's job of monitoring citizen behavior is simplified because the police have to focus on only illegal behavior. Of course even the set of all illegal activity is too fine-grain for the police to capture every instance. Not all offenders get caught.

The police-criminal interaction is high-frequency because police need to respond quickly and at a level congruent with the behavior of the perpetrator. Police pursuits may involve very fast dynamics, with both the police and the suspect behaving very quickly. Sometimes high-speed chases do occur in the manner of Hollywood action sequences. These fast, high-frequency dynamics are in stark contrast to the dynamics of the court. Once the accused is caught, the process becomes a matter of waiting.

The next step in the legal process is appearance before the state courts. The state level can be considered an intermediate level between the superordinate context of the supreme court, and the lower-level, high-frequency functions of police law enforcement. In the state courts cases are processed slowly but surely. Judges are the memory in the system, with state judges serving a moderate but fixed term before reelection. Judges are responsible for keeping justice consistent over time. The judge's education and past experience determine which elements in a case are considered relevant violations of the law, and which should receive more weight than others. This information is integrated, and a decision is made.

The state judge is operating at a lower frequency than the police officer. The judge's decisions incorporate not only local social issues but also the judge's education and values, which invoke legal debates argued in many countries over hundreds of years. The judge's relevant data base, which is brought to bear in every decision, is more extended temporally and spatially than the knowledge required by an individual police officer to bring the accused to justice.

The highest level of the American legal system is the Supreme Court and Constitution. There are many levels of appeal between the local judge and the Supreme Court, but these are details in the present argument. The highest level serves as the context for the daily workings of the entire legal system. A stable context is crucial for continued and effective functioning of local communities, and so it is no coincidence that the term of office for Supreme Court and Federal Justices is life. Furthermore, the Supreme Court uses the intent of the original creators of the Constitution as one of its central guiding principles. Any Constitutional amendment requires a two-thirds vote in the Senate and is undertaken only in dire cases. The Constitution and Supreme Court justices, therefore, serve as the ultimate memory in the system.

A low turnover rate of Justices and difficulty amending the Constitution insulates the highest level from the influence of local changes in public opinion. Changing the upper level is always a last resort. The local judge, in contrast, looks more often to local precedence and can be swayed by public opinion, especially in an election year. However, any ruling at the State level which may violate the Constitution can be pursued at the level of the Supreme Court, where its consistency with the Constitution is scrutinized.

It was with great wisdom that the founders of the United States invented a system that embodied a low-frequency, stable context. Notice how the Supreme Court and Constitution constrain the lower levels, not by being energetic and active but by preserving tradition and allowing change to occur only infrequently. The design of the system prevents attempts to make the highest levels flexible and sensitive to high-frequency signals, like fads or election gimmicks. Frequent changes of the upper level threatens the stability of the entire society. In the final analysis, a stable, questionably just legal system is better than a system characterized by a series of rapidly formed and dissolved legal frameworks. Systems where revolutions and dictators change the core political systems of the country are the best argument for a stable legal system.

FIGURE 5.2 "The army *consists* of the soldiers of all ranks, and it *contains* them all."

Photograph by Paula Lerner

Nestedness

The example above deals with empirical, *non-nested hierarchies*. In non-nested systems the general ordering priciples can be seen in terms of four characteristics of upper levels relative to lower levels. Upper-level entities 1) behave at relatively low frequencies, 2) behave with less integrity, 3) offer context, and therefore 4) constrain lower entities. *Nested hierarchies*, on the other hand, are those with the added requirement of *containment* of lower levels by upper levels. Although nesting is just one more principle, nested hierarchies are more readily recognized as being hierarchical than most others. For example the hierarchical order of cells in tissues, in organs, and in organisms is very obvious. An army is a nested hierarchy. The army is the context of all the soldiers, and it exhibits lower-frequency behavior than do its subunits. It offers constraints over all its personnel, from the privates to the general. The entire army has less integrity than its subunits. However, none of this makes it a nested hierarchy. The critical characteristic that makes it a nested hierarchy is that the army *consists of* the soldiers of all ranks, and it *contains* them all. By contrast, the military command is a significantly non-nested system. The general is the head of a military command hierarchy, making utterances and decisions at a lower frequency than his troops. The general is the context of the troops, and is the highest member of the command hierarchy. Even so, he does not consist of his solders, and does not contain them in his physical being. Therefore there is an important non-nestedness to the command structure. The army itself, on the other hand, can be validly seen as consisting of all the soldiers of all ranks, and can be described as a nested structure.

Pecking orders are an example of a control hierarchy that exhibits all the principles of hierarchical order except containment and bond strength. The top dog constrains and is the context of the other lower members, but it does not physically contain, nor does it consist of, the other members of the hierarchy. Compared to nested hierarchies, non-nested hierarchies have a converse set of strengths and weaknesses, and so they perfectly complement the nested conception.

The tangibility of nesting leads to much more than just tractability and convenience. Unlike non-nested systems, nested systems are determinable from a knowledge of the parts. For example, the position of the army and its aggregate behavior can be determined from a full knowledge of the whereabouts and activity of all the soldiers. On the other hand, the behavior of the general is not determinable from a knowledge of his command because there are other degrees of freedom that exist independent of the army of soldiers under his command. These degrees of freedom could be the general's individual military training, experience, and predilections, none of which are determinable from a knowledge of the army he commands at the moment. It emerges that nested hierarchies are particularly helpful at certain stages in developing understanding. However, the sword is double-edged. Nested hierarchies can also be obstructions to understanding if levels are reified.

The Robustness of Nesting

The central organizing principle of nested hierarchies is containment. Although containment is the special principle in nested systems, it is not usually the organizing criterion that precipitates an investigation. For example, medicine is not particularly interested in the fact that the diseased person happens to be nested; the nestedness of the viscera in the abdominal cavity is taken for granted. Some other explanatory criterion is usually the one of interest, such as the functioning of antibiotics, and nestedness serves only to draw attention to the question of interest in the first place.

Nestedness is an important ordering principle that allows for an accommodation of the richness of biological and social systems. The criteria that link successive pairs of levels can shift, but the system remains conceptually cohesive and tractable because nesting and containment keep things straight. For this reason nestedness is very valuable, but its power comes at the cost of constantly shifting between significant criteria. The chameleon character of nested systems is an invitation to err by applying criteria that work at one level to other levels where they do not. The mistake is to assert that criteria that apply at one level necessarily apply to others. In fact, only nestedness applies across all levels of a nested hierarchy. Ordering criteria are free to change, while nesting holds the conceptual system together.

Explanations of one level need not have anything to do with explanations of other levels. For example, despite the unequivocal success of biochemical analysis of biological material, the relationships of those mechanisms to other levels in biology are often tenuous. The links between phenomena at the level of whole organisms and biochemistry are usually specific to the path of reduction that stimulated the line of research in the first place. We know quite a lot about biochemistry, and it has become a coherent body of knowledge, linking specific processes in organisms from fruit ripening to cancer. However, the links are made from a very specific organismal phenomenon to a very particular part of biochemistry.

A general linking of most organismal behavior to biochemistry or of most biochemistry to organismal behavior is far from being realized. Although many specific and valuable links will be achieved in the future, a generalized linking of organisms to biochemistry is not possible because as one moves beyond considering a single individual to interactions between individuals, biochemistry is not the appropriate level of explanation. Of course all organisms function on the basis of biochemistry, but casting organism behavior and interactions in biochemical terms is impossibly complicated. For example, consider the outcome of competition between two plants, such that one comes to set seed and the other does not. Each separate occurrence of that organismal phenomenon is unique in the biochemistry that it invokes, and the details of the biochemistry are irrelevant to the eventual outcome.

Despite the unbridled optimism of its practitioners, biochemistry will not explain great tracts of biology because it cannot. Above the level of organisms, the situation is even more hopeless. Here, entirely different explanations, like food webs, speciation, pair bonding, and other concepts that are explicitly not biochemical are the rule. Just because biochemistry is nested inside organisms this does not mean that higher levels in the nested hierarchy of life have the same type of relationship to their lower explanatory levels as do biochemical pathways to enzymatic reactions.

The intuitive power of nestedness allows it to be taken for granted, allowing other criteria that may coincide with containment to be considered. These additional criteria may run parallel to containment for only one pair of levels, but the parallel entities provide points of articulation that link between pairs of levels. For example, genetic relatedness between individuals links most members of a nuclear family. Nevertheless, nuclear families are also economic units as much as they are the consequence of parents having children. The economic relationships of the nuclear family, as it ties into society at large, will have little to do with genetics. One does business with people who are not relatives. As one shifts between levels the relationship that links levels is free to change. Furthermore, the scalar changes that come with upper levels

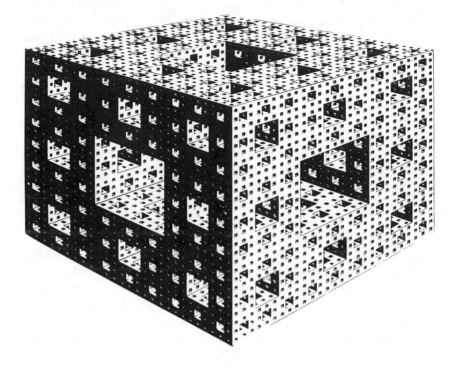

FIGURE 5.3 The Menge sponge results from a regular excavation of a cube, in a continuing process, down to infinitely small scales. At that point, it has infinite surface area, and no volume. Fractal structures are counterintuitive. It is clearly less than fully three-dimensional but also obviously more than two-dimensional. It has a dimensionality between two and three.
After Benoir Mandlebrot, with permission

containing lower levels may force a change in type of relationship. Principles that apply to small-scale entities may not be able to operate across the span of the higher level, and so a change of criteria is not only possible but necessary.

There is one special case where a single relation holds across changes in scale, and only recently has science had an adequate means of dealing with these peculiar nested systems. They are fractal, and require a distinctive, fractal geometry. Clouds are an example. Small and large clouds look alike independent of scale. The outline of the whole cloud looks like a small section of its own outline, seen through a telescope. The reason for this scale-invariant phenomenon, called self-similarity, is that the mechanics that link big clouds to medium-sized clouds and medium clouds to small clouds are the same at all scales. Another common system that repeats patterns at large and small scales is turbulence in a fluid. Large eddies have small eddies coming off them, and these in turn have smaller eddies coming off them. The very same equations describe the relationship between eddies of a given size and those of both the next size up or down. In nested fractal systems the relation that links between scalar levels is constant, but these cases are the exception rather than the rule.

While there are material systems that exhibit self similarity, some of the most striking are the creations of mathematicians. The Menge sponge is made by a theoretical process of taking a cube and punching holes through it in all three dimensions (figure 5.3). Imagine a three-by-three grid giving nine squares, as in a tic-tac-toe board. If the board is a side of the cube of the Menge sponge, then the hole is punched through the center square. This of course leaves eight squares surrounding the hole. Divide them into three-by-three grids, and punch a hole in the middle of them. Around each of these smaller holes there are eight yet smaller squares. Punch holes in them, and so on. If this process is done on all six faces of the cube, and is continued to smaller and smaller squares, the sponge has the remarkable property of zero volume and infinite surface area. Thus the sponge is larger than two dimensions, but smaller than three. In fractal geometry things can exist in between whole numbers of dimensions.

The usefulness of fractals is that they can deal with generalized patterns across widely differing hierarchical levels. It is a mathematics that can deal with a whole class of apparent contradictions. For example, from a great distance a field naturalist can see enough similarity between elm trees, which are themselves all different, so as to unerringly identify elms from oaks. All elm trees have a particular fractal structure that overrides the differences in their individual details. Likewise, oak trees share a different fractal structure that belongs to their species. In both cases fractal geometry offers a parsimonious explanation as to how the naturalist achieves accuracy in identification from great distances. The message from fractal geometry for nestedness is that for the most part nested systems only rarely invite using the same criterion to link levels through the entire span of the hierarchy. If it does appear sensible to use the same criterion from top to bottom, then one is dealing with a special case, a fractal system. Explanation in most nested systems requires supplying criteria beyond obvious containment.

Nested versus Non-Nested Criteria

We have just seen that flexibility in changing criteria as one changes levels is required when analyzing nested as opposed to non-nested hierarchical systems. There is an intuition among biologists that non-nested hierarchies are somehow more abstract and less generally hierarchical than nested hierarchies. This may be due to the fact that containment is a concrete relationship. In contrast, the integrity of non-nested systems more often derives from abstract criteria, such as control and behavioral frequency. In fact, however, it is nested hierarchies that are the special case, for they impose the additional criterion of containment. While it is usually not appropriate to link all levels in a nested system by the same criterion, in non-nested systems it is not only possible but desirable.

In non-nested hierarchies containment is not there to anchor the conceptual scheme. Hierarchical levels are therefore more flexibly ordered in non-nested hierarchies. Such flexibility invites confusion unless one keeps the same ordering criterion from top to bottom. The top level may be top because it is the slowest-moving, most intransigent, or most generally constraining. Take, for example, an ecological food chain as an example of a non-nested hierarchy. The ordering criterion is predation, and the highest-level carnivore is the top of the hierarchy who constrains all, and is preyed upon by no one.

An alternative food web ordering is according to food dependency and energy flow. Here the top carnivore is on the bottom, dependent on the constraints imposed by the productivity of the lower levels. For example, an aquatic ecological system could be described as controlled from the top by the top carnivore. The recent clarification of the water in Lake Michigan comes from the introduction of big fish that suppress little fish. With the little fish gone, small animals are released from fish predation and can then crop down the floating plants, which in turn leads to clear water (figure 5.4). An alternative example is Lake Erie, where improving water quality came from reducing the amount of phosphorus in the water, which suppressed the floating microscopic plants. The animals in the system are all constrained by the quantities of microscopic plants.

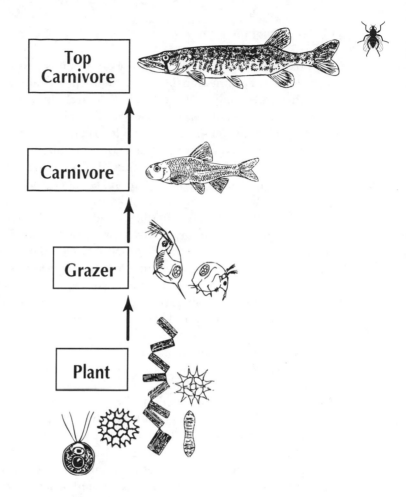

FIGURE 5.4 "The recent clarification of the water in Lake Michigan comes from the introduction of big fish that suppress little fish. With the little fish gone, small animals are released from fish predation and can then crop down the floating plants, which in turn leads to clear water."
Drawn by Kandis Elliot

This energy-dependent, non-nested hierarchy has the small plants rather than the top carnivore at the top. The critical point here is that in all cases one explicit criterion unites all levels and gives the system integrity.

Because non-nested hierarchies apply a single criterion from top to bottom, units of measurement apply across all levels. Common units facilitate comparing between hierarchies that use the same type of unifying relationship. In food chains, for example, it is possible to use average efficiency of calorie transfer across all levels to characterize an entire hierarchy, and compare food chain hierarchies with one another. Comparing between hierarchies is less available in nested systems. In the nested case it is uncommon for a single criterion to apply across more than a few levels, and this causes each conception to be unique.

Earlier we identified that research and formal investigation has two phases. The first is exploratory, when one is not committed to particular entities. New structures are posited as one stumbles across different aspects of the material system. In the second phase, which is more formal, the critical questions emerge in a framework that, if not firm, is identifiable. Decisions as to important entities and distinctions have been made, and the power of the constructed intellectual framework is tested.

Nested hierarchies are most suitable for exploration, where nestedness draws attention to a hierarchical system and helps keep order while intuition explores some poorly known phenomenon. For example, the fact that the human body contains organs is undeniable and obvious from the outset. After the exploratory phase, however, nestedness itself becomes uninteresting, or at least incidental. With further elaboration nestedness becomes subordinate to new, less general organizing criteria, such as rates of enzymatic reactions. The new criteria link between adjacent levels, but do not apply from top to bottom. More often than not, nesting serves its purpose as the point of departure for the investigation, only to be abandoned as other ordering relations come to the forefront.

Non-nested schemes, on the other hand, are less appropriate when exploring unknown situations. In non-nested schemes the criterion is elaborated first, and then material examples that fit the criterion are sought. For example, first predation is considered an interesting organizing principle, and then it is applied to aquatic and terrestrial examples. The advantage of non-nested hierarchies is that they are very precise, and state explicitly which principle is the central idea in the investigation. The very inflexibility that denies non-nested schemes in the exploratory phase provides a firm framework that presses across the entire span of hierarchical levels.

Conclusion

Hierarchies can take many forms and use a variety of principles to order the levels. With some ingenuity it is usually possible to translate between different generalized criteria. For example, ordering levels according to constraint can also be achieved by relative frequencies. Indeed, it is exactly low-frequency behavior that lets constraints hold the lower-level entities in context, so that their behavior is predictable. Nested systems are distinctive in that the criterion of containment does not necessarily apply to other types of hierarchical order. Nested systems do manifest all the other major ordering principles, but not all systems show nestedness.

Hierarchical orders serve particular purposes. The ordering principle that is employed is chosen for reasons of clarity of expression. There is no better principle than parsimonious insight. The contribution of the investigator is to cleave nature with efficient perspective that best suits the purpose at hand. In this chapter we focused on the ordering principles of constraint and nestedness. In the next two chapters we examine ordering levels in terms of filters and surfaces.

6

Filtering Information

We have identified the important role of scaling in observation and have also probed ways to use relations to order levels. We now turn to issues of informational exchanges between levels and sources of discontinuities in observation. One reason that levels have integrity is the free flow of information within entities at levels, contrasted with obstructed or delayed flows between levels. This chapter looks at material and informational flows that give levels their identity and integrity. The patterns of these flows separate levels from other levels, and we see this as we turn to address discontinuities. Some discontinuities originate in the observed, while others emanate from the observer.

Earlier we saw that one heuristic for ordering hierarchical levels is according to the spatial and temporal characteristic frequencies of empirical entities. An alternative conception is to think of levels in terms of varying rates of information flow. The appropriate notion here is the filter. In the present chapter we discuss filters and discontinuities, and in the next chapter focus on issues of interactions between parts as an additional source of level identity and integrity.

Filters fix the rate of information exchange. Some filters are very tangible, like the pieces of porous paper used in making coffee. Other filters are equally common but more abstract, like filters in an amplifier that remove background hiss in a recording. Whether tangible or abstract, the concept of filters will help operationalize our ideas on empirical entities and hierarchical levels.

FIGURE 6.1 "Nothing reacts immediately." She is ready to dance; he is hesitant. *Photograph by Paula Lerner*

The Nature of Filters

In an observation, it is helpful to think of filters in at least two separate places. One is found in the rate of information flowing from the observed. The other is a filter associated with the observer input channels. On the observer end, both grain and extent contribute to the input filters. In addition to filters that occur in the process of data collection, it is also helpful to interpret aspects of the material system as filters. Three aspects of filters that are relevant to hierarchical models are 1) filters as a means of *attenuating* signal; 2) filters as they *delay* signal; and 3) filters as they *integrate* or average signals. These three aspects of filters can apply to filters both in the material system and in the observation protocol, but we will consider them most explicitly in the filters of the material system.

Filters Attenuating and Delaying Signal

Nothing reacts immediately (figure 6.1). There is always some sort of interpretation before a response is made. A stimulated entity may be divided into three separate parts: the *input channel*, the *internal system* itself, and the *output channel*. One way to look at the internal system is as a filter that connects input to output. Filters delay output by introducing a lag. The *lag* is the difference between the time the input signal impinges on the surface of the entity and the time the output signal or response leaves the entity. Filters may also modify input before it becomes output by integrating or weighing some aspects more than others.

By mediating between input and output, the entity as a filter can cause apparent discontinuities in the observed by limiting or slowing the passage of energy, matter, or information. In wildflowers, for example, differences between species can be described as the product of a filter attenuating genetic signal. Think of a species as a pool of genetic material that flows readily, by sexual union, through individuals to successive generations. The reason members of the same species look more or less the same is that breeding occurs freely within the species. Members of different species, on the other hand, look different because genetic exchanges are inhibited. All members of the common English primrose have many genes in common because they all belong to a large population, with many recent shared ancestors. The same applies to the English cowslip (figure 6.2). Within both species there is an unimpeded flow of

FIGURE 6.2 The cowslip (above) is common in English meadows. The primrose is a woodland species. They hybridize perfectly but still function as separate species. *Drawn by Kandis Elliot*

genetic material, and individuals who share a genetic pool are closely related.

Interspecies hybrids are often infertile, as when a horse and donkey cross to make a mule. Hybrid sterility is not, however, universal. In the case of the cowslip and primrose hybrids are fully fertile, meaning that there is no intrinsic genetic impediment to breeding. Hybrids can breed freely with members of either parent species to form intermediates of three-quarters primrose to one-quarter cowslip, and so on. It should be no surprise that, given a full fertility of hybrids, the two species do indeed blend into each other in nature.

Given the full fertility of hybrids, one might ask how the physical form of members of the species retains any identity at all. Why do the phenotypes of the two species not converge on some average condition? The answer is that the primrose is well adapted to woodlands, and occurs there almost exclusively. The cowslip, on the other hand, belongs in pastures. Therefore, when it comes to breeding, a primrose has greater access to others of its own kind than to cowslips, and vice versa. The difficulty for the hybrids is not an incapacity to breed but rather that there is no intermediate habitat between woods and fields in which the hybrids can perform well enough to become strong, vigorous plants. Given a habitat that is functionally intermediate, the hybrids could breed with the same vigor as members of either parent species. Lacking that habitat, hybrids are less successful than their parents, and produce fewer offspring on average. There is, therefore, a more rapid exchange of genes inside each species than between species, and the low success rate of the hybrids serves to attenuate the flow of genetic information between species.

All this translates directly into terms of filters. Any gene passing from one species to another must pass through a hybrid filter to get there. Genes moving across to the other species are blocked by early death, or inept reproductive performance in the hybrid individuals in which they reside. This hybrid filter on genetic flow maintains the interspecies boundary. There is some exchange of genes between species, but that exchange has to pass through a filter of hybrid individuals who attenuate the passage of genetic signal. The result is an implicit boundary, or surface, between species.

The flow of genetic material inside each species is relatively fast, and it is that speed of communication that keeps all the members of each respective species looking similar. Any mutation occurring in one species, on average, reaches members of the other species only after a delay. Furthermore, during that delay, many of the mutations are lost, so the mutation signal attenuates across the divide. The attenuation is imposed by natural selection on the transfer of genetic information.

Filters Integrating Signal

Related to the notions of signal attenuation and delay is the concept of *integration,* or averaging. Integrating input may be a simple averaging of portions of the signal string. For example, two numbers come into the filter, say 1 and 3, and one number comes out, 2, which is the average of the input. The output is a single number, corresponding to a string of two input numbers. Lag relates to integration, much as it does to signal attenuation. A filter with a longer lag might remember three numbers at a time, and take the average of them. The longer lag comes from the extra time it takes to read three rather than two numbers. The integrated output signal is a smooth response to an input signal. The smoothing occurs because adjacent outputs share at least two input numbers.

In most filters, some parts of the input signal string are treated as more important than others. In the case of the human eye, for example, light is more important than dark. In human vision dark is the absence of light, not an active state in its own right. Seeing light comes from light bending photoreceptive molecules on the retina. A delay in light detection comes from the time it takes a receptor molecule to straighten out again. As long as the molecule is bent, the eye considers that the world is light. If a short period of dark intervenes but gives way to light before the receptor molecule can straighten, then the flicker of darkness passes unnoticed. It is on this principle that movies operate. The frame changes every twenty-fourth of a second, but the eye's limit on detecting darkness is fifteen times a second. The picture therefore appears continually bright. The human eye works with a window detecting light over a relatively long time, a fifteenth of a second. Accordingly, the important input stream passes through a filter, the eye, and influences appearance disproportionately in the output, the sensory experience.

Flicker frequency is the visual system's limit on detecting periods of darkness. For other animals the world seems very different from the way it does to us, because their flicker frequency is different from ours. Not

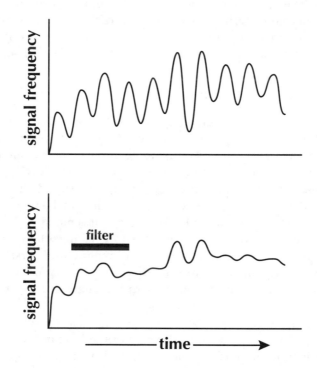

FIGURE 6.3 Two passes through the same data. The first pass is raw data over time. The lower graph is the same data read over a smoothing window that averages across a hump and a valley in the raw data. The effect is to smooth the spikes out of the original data. Notice that almost all variation disappears, except the big trough in the middle, and even that is smoothed out.

only can a fly in a cinema detect the movement of individual frames, but it can also detect the effect of reversing alternating current that supplies electric lights. Household lighting runs on current that alternates fifty to sixty times a second. Table lamps, therefore, flash off and on at that rate. In a sitting room, the fly might wonder why, since we are responsible for the lighting arrangements, do we not do a better job. Although we cannot see anything but a steady light source, the fly with its high flicker frequency sees the room as alternately light then darkening.

The physical size of a fly relates to its high flicker frequency. It is no accident that the fly is small and can detect higher-frequency recurrences. The nerves in a fly are necessarily short and so can transmit impulses end to end relatively fast. As far as a fly is concerned, this is not just convenient; it is essential to survival. The smallness of a fly makes its experience of local circumstances in our world relatively large. In the world of a fly, a millimeter can be a very significant distance, and something moving half a centimeter could be cataclysmic. The fly uses a filter with only a short time lag, and so integrates experience over small spaces and short times. The visual experience of the fly is at a much higher frequency than our own visual experience.

As a contrast to the fly's perception of light, consider what happens when ecologists are interested in measuring the functional environment of plants. Moment-to-moment human perception of light contains too much variation to be of much help in assessing the light regime of a plant. Accordingly, ecologists sometimes measure light chemically over an extended period. The amount of a light-sensitive chemical destroyed over the entire day is the measurement. If we think of the measuring device as a filter, a short period of intense sunlight on an otherwise cloudy day has the effect of only slightly increasing the amount of chemical destroyed. In forests, the strong light often reaches the forest floor only as little patches of light called sun flecks, as the sun moves for a few minutes across a small opening between branches. Using a chemical measure, a sun fleck only comes through the filter as a small increase of light, on average, for that day. The spike of light is smoothed to a low, flat value belonging to the entire day. The length of time the light sensitive chemical is exposed gives the length of the window. The longer the exposure, the more the bright sunny spot merges into a measurement period of a generally dull day (figure 6.3).

Filters attenuate, delay, or integrate input signal. There are explicit, mathematically defined relationships between these three properties, but that is more pertinent to engineering and physics than to our overview presented here. In addition, filters use windows of varying sizes. A filter that considers signal from the distant past has a long time window, and is relatively unresponsive to short term happenings. Recent inputs to the filter are averaged out against information already in the window. A short-term window, on the other hand, is highly responsive, but at the cost of long-term memory. Inputs are quickly converted into outputs, and signals from long ago are forgotten. Filters are important because they link and separate entities and levels. High-level, low-frequency empirical entities typically operate with long-term windows, and low-level, high-frequency entities typically operate with short-term windows.

The Role of Filters Between Levels

In chapter 3 we argued that upper levels of organization are populated by larger, slower-moving entities, and serve as the context for lower levels. To translate the notions of characteristic frequency and constraint into terms of filters, it is instructive to think how the fly perceives us, and we the fly. At the scale of commonplace human experience, fellow humans appear active, often moving, sometimes quite fast. For the fly, the human is usually nothing but a firm, passive substrate on which it sits. From time to time that substrate moves slowly, whereupon the fly might decide to take flight. Slow as our movement is to the fly, it might be the manifestation of our desire to be rid of the pesky thing. For a fly, your fastest movement is so slow that its course can be predicted easily, and the insect avoids the moving hand. Without the leverage offered by a fly-swatter, to speed up the effect of wrist action, humans cannot behave at a speed that presents a problem to the fly.

For a mosquito, on the other hand, a human being is a passive resource. Mosquitoes find us by moving up a heat and carbon dioxide gradient. At the source, the mosquito finds the bag of food, us. So different is the world of a mosquito that we are too big for it to see, and it is on that principle that insect repellents work. Repellents do not actually repel, they only confuse the guidance system. Without a guidance system, the insect flies straight past, unaware that a food source is nearby. We occupy only a very small part of the spatial volume around a picnic table, and mosquitoes without guidance only bump into a victim occasionally. If they do, they will bite, and that is why heavy infestations can still find us, albeit by statistical average.

For both the fly and the mosquito we are the environment. *Environment* can be seen as everything the system encounters that behaves more slowly than it does. Speed of behavior, be it slow or fast, is related to filters. Slow behavior follows from an entity employing long windows in its filters. A long window means that signal is read over a relatively extended time period, before it is processed. In general, long windows process signal by averaging, which smooths spikes of local events. The effect of averaging over a long time period is to flatten out blips in

FIGURE 6.4 In killing a fly, a creature that lives in a very differently scaled world, there are several strategies, all of which involve rescaling. The flyswatter rescales the human to function as quickly as the fly. Strategy one: get into the same realm and win. Strategy two: the horse does not expect to kill the fly, but it does offer a constantly unfavorable environment. Strategy three: with the bug spray we become the ultimate context. If bug spray killed us, we could not do it. However, it makes the fly sicker faster, so we fill the insect's entire space with poison. It is hard to play on the fly's terms, so we manipulate its context.

the signal, so that even strong but transient episodes pass through the filter as only gradual increases in signal amplitude.

What constitutes active behavior at an upper level of organization is a constant for lower levels. Being the context, upper levels constrain lower levels in a general way. Where the human sits determines where in the room the fly sitting on that person will be. Unless the fly moves to something else, it has no say in the matter. Despite these general constraints, the context does not have explicit control at the lower level. This is because the upper level cannot be sufficiently responsive to the details of lower-level behavior. The fly moves before you can hit it with your bare hand. The upper level can have only a general, not a specific, effect. A horse presents a rump that is constantly whisked by its tail, and anything less than constant whisking will not keep the flies off. To combat insects, we sometimes change the entire room into a space full of bug spray. Bug spray is effective because it works on all flies in the space, independent of the details of their individual behaviors. Constancy rather than short-term action is the tool of entities at upper levels (figure 6.4).

In addition to being a constant context, there is a second way that things which behave slowly constrain high-frequency behavior. Even if a fast entity gives a signal that could affect the upper level, the slower entity takes its time to read the signal. All the fast entity's rapid jumping up and down is fed into a filter with a long window, before it is read by the slow entity. The fast entity can do nothing but wait for a response in the context of the slow filter. This can explain how many social situations function, and how social control is executed. For example, power in an administrative system is exercised by offering only periodic utterances and ignoring fine-grain details. Faculty must wait for a decision from deans, and deans are relatively uninformed about day-to-day faculty activities. Part of the success and power of Howard Hughes, the aviation magnate, came from his pathological inability to make decisions. His competition was therefore forced to make a move and, in so doing, boxed themselves in and lost their options. At that time Hughes's best decision was so obvious that it was hardly a decision at all, and even he could make it. Thus one source of power and control is a matter not of making active decisions but of making lower levels wait for decisions. President Carter deserves more credit for what he achieved, but a fair criticism of his Presidency is that he laid out too many of the details and was too responsive. By contrast, the unlikely success of President Reagan came from his simplistic presentations that ignored the cogent arguments about the details being presented by his opposition.

Similarly, in the legal system higher levels work with a longer window and slower filters than do lower levels. Judges are an example of a slow, integrating filter that weighs some parts of the input signal more than others. In instructions to a jury, and in taking only some evidence as admissible, the judge operates exactly as a filter. Some signal is blocked, and other parts are allowed to pass. When there is no jury, the judge listens to all the arguments in the case and gives a single utterance as the output: guilty or innocent. The evidence is the input, and the decision is the output of the judicial filter.

The judge also operates as a higher level of organization with respect to law enforcement officers. No matter how guilty the arresting officer may feel the accused to be, it is the judge who says that the alleged perpetrator is guilty or innocent, or even that there was a wrongful arrest. Note that while the police may spend time building a web of evidence for the prosecutor, the judge's decision draws upon an even longer legal history. The police filters focus on fine details, and are only part of the input to the judge. Other considerations, like the testimony of social workers, psychiatrists, or legal precedence from other decisions all contribute to the final "guilty" or "innocent" decision. The judge's decisions supersede the expectations of the police in any given case.

Military command is another example of higher levels operating with coarser filters than lower levels. Arthur Koestler in *The Ghost in the Machine* pointed out that when a general asks for information about enemy troop strength and position, the orders take on ever finer-grain character as they go down the chain of command. Eventually, the original request becomes Sapper Smith at the end of the trench putting up his head to see what is happening fifty yards away in enemy trenches. Then, in the reporting process, the filters work in the other direction, so that factual details about the small front section seen by Smith are interpreted in an ever widening context as they are sent up the chain of command.

As the information passes to higher levels, it is integrated with observations from other sappers at their positions on the front. Eventually, the general finds that there is a concentration of opposing troops at the western end of the line. All the details of the individual reports are lost, but an integration of them provides the big picture. Note that the individual soldiers have no idea of the big picture. The individual soldier's filter is

too narrow to see the broad pattern (figure 6.5). The filter is too fine-grain and his extent too narrow. It is for that reason that all the great generals, since Alexander the Great, have hung back so that they can take the big picture into account. Alexander was the last to lead his troops literally into the attack. Thus his army was incapable of cohesive response, and won every battle because of enormously superior set formations of staggered phalanxes and staggered lines of spears going into the battle. Caesar fought with his troops on the front to inspire them, even sending away the officers' horses, so that the men knew that their fate was his. However, Caesar did this only at times of crisis, having surveyed the entire field and identifying where he needed to be. In contrast to the individual soldier, the modern general operates with a filter of coarse grain and wide extent.

Continuous versus Discontinuous Change

The notion of filters is helpful in understanding the issue of continuity versus discontinuity. Grain and extent, both aspects of filtering information, have significant roles to play in the emergence of apparent discontinuity. An exploration of these matters comprises this final section of this chapter.

Once empirical entities are delimited, the issue of describing changes naturally arises. One common question that arises in changes in a phenomenon over time is whether the disjunctions observed are truly continuous or truly discontinuous. Each major discipline has been concerned with this issue for its particular subject matter. For example, is the variation of vegetation in ecology continuous or discontinuous? Are social classes discrete in sociology? Do changes between cultures in the archeological record indicate a discontinuity between cultures? Is recession a discrete phase in the economic cycle, or does it have to be big enough to be called a depression before it indicates a discrete phenomenon? As often as not, there are underlying realist agendas that turn on

FIGURE 6.5 The soldier does not know. A recruitment poster of the day encourages the young men to look to the truths they can tell. And yet when they are really at the front, they do not know much of what they do then, let alone what it might reflect later. In retrospect, the face of the man looks haunted—surely not what the distributors of the poster had in mind. In the heat of battle, the soldiers do not know.

FIGURE 6.6 "To the great grandmother, who sees her great grandchild only once a year . . ."
Photograph by Paula Lerner

the argument that discrete means something is real, whereas continuity suggests a human artifact.

Discontinuity has one set of implications in the material system, and another as to observational decisions. With regard to the material system, there must be a certain amount of material heterogeneity for discontinuity to appear. The observer's perception of discontinuity depends, to an extent, upon there being some material change. However, material change alone is not sufficient. The coherence of our experience would indicate that, even across change, there is some continuity moment to moment. Nevertheless, greater change invites a decision that the change is discontinuous. With all else equal, the greater and faster the change in the material system, the easier it is to experience that change as discontinuous.

With regard to the observation protocol, the wider the extent, the more likely it is that one will encounter some change that is large enough to warrant calling it discontinuous. On the other hand, in a regime with narrow extent, small changes that would be overlooked in a larger universe might stand out enough to warrant being called discontinuous. Thus both wide and narrow extent can lead to an apparent discontinuity. We will give specific examples later.

Grain also plays a complicated role in assessments of continuity. A coarse grain can miss sharp but local differences by sampling either side of the local change, so that it does not appear in the data set. The spike of change in the material system is averaged with the other, more normal, conditions that also contribute to the data point. In this way, a coarse-grained sampling regime can indicate continuity where a finer-grained measurement regime would detect impressive change. Alternatively, a fine-grain measurement regime might catch intermediates between the states found in a coarse-grained sampling regime. Thus an apparently discontinuous change under coarse grain might emerge as continuous when it is dissected with a finer-grain observation protocol. Therefore, a wider and narrower extent, and coarser and finer grain, both influence the perception of discontinuity.

To illustrate the relationship of grain and extent to apparent discontinuity, consider child development. Apparent discontinuities in language development may arise depending on whether observations are made once a day, once a month, or once a year. The transition from babbling to sentences can appear either continuous or discontinuous depending on the measurement regime. To the great grandmother who

FIGURE 6.7 "An example of a discontinuous change is the difference between a pupa, a caterpillar, and a butterfly." *Drawn by Kandis Elliot*

sees her great grandchild only once a year, it may appear as if the child has undergone a radical, discontinuous fluctuation. This is because the time between visits spans critical phases of development. In contrast, to the parents who see the child every day, the transition from practicing phonemes, to single words, to word combinations will appear entirely gradual, despite a spurt in development here and there. Discontinuity seen with a coarse grain over a long time can easily appear continuous on closer inspection with a fine grain. Apparent discontinuity over the long run is seen daily as a set of small increments that amount to a continuous change. Therefore temporal grain greatly affects appearances by influencing the level of analysis.

On graphed data, plots of steps and plateaus may be highlighted as evidence that changes in a phenomenon are discontinuous, or stagelike. Conversely, smooth curves that rise gradually are taken as evidence for continuous change. An example of a discontinuous change is the difference between a pupa, caterpillar, and butterfly (figure 6.7). These three stages correspond to qualitative shifts in constraints upon mobility. The crawling caterpillar is restricted to two dimensions, the pupa has no mobility, and the flying butterfly moves in three dimensions. These different sets of constraints can be considered three qualitatively different stages in the life cycle. By contrast, an example of continuous change is the growth from a hatchling fish to adult size. Here, change in mass is gradual and incremental. A smoothly increasing mass might be taken as evidence that a quantitative, continuous accumulation has occurred. In the case of quantitative change, the same constraints apply, and there is simply more of the same "stuff."

Monthly or annual sampling of a growth sequence ignores all information about daily fluctuations, and allows growth spurts to emerge as discrete steps. Daily sampling, on the other hand, allows essential continuity to emerge, and spurts of growth appear as nothing more than accelerations in continuous change. Neither the monthly nor the daily growth curve plot answers the question whether change in a phenomenon is quantitative and incremental, or discontinuous and stagelike. Rather than ask whether a phenomenon really is continuous or discontinuous, a more profitable question is cast in terms of changes in constraint. Ask whether a continuous description fails to capture interesting changes in constraint. Alternatively, ask whether a stage description fails to identify the many precursors that provide the foundation for that which, from the vantage point of a coarse-grain measurement scheme,

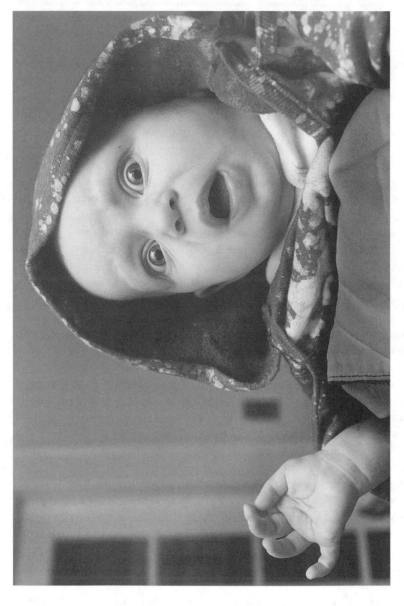

FIGURE 6.8 "Without language and symbols, the infant's ability to remember the past is extremely restricted." *Photograph by Paula Lerner*

appears as discontinuous change. The question is not "What is nature, really?" but rather, "What is the most efficient and useful description of nature, given the question at hand?" Empirical data alone will not resolve issues of fundamental discreteness or otherwise, regardless of the form of the data plot. Questions of continuity versus discontinuity are a matter of measurement and observer emphasis on stability or change of constraint.

New terms or classification categories are introduced to capture stage-like breaks, which may, nevertheless, be the product of an underlying, continuous process. For example, a species name implies a classification, but this does not require the absence of intermediates between species. Indeed, there are many examples of hybrids, as between the cowslip and the primrose that we analyzed earlier. The use of their Latin species names, *Primula veris* and *Primula vulgaris*, indicates only that there are fewer hybrid intermediates in nature than members typical of the respective species. A continuous mechanism, gene flow, produces phenotypes sufficiently divergent to demand a qualitative class distinction.

The fact that measurement greatly influences the form of data does not preclude emphasizing either continuity of discontinuity in an underlying theory. The important issue is whether the model of change is helpful and predictive. Often both continuity and disjunctions need to be invoked. For example, while individual connections between nerves are relatively discrete events, increasing synaptic connections is a continuous process which underlies increasing cognitive complexity. At the higher level of emergent behavior, the difference between a presymbolic, prelinguistic infant and a symbol- and language-using infant represents a qualitative shift in constraints. Without language and symbols, the infant's ability to remember the past is extremely restricted. While a common underlying mechanism, increasing synaptic connections, can be described as continuous and quantitative, the emergent behavior is more profitably described in terms of qualitative stages and shifts in constraint.

Arguments emphasizing either continuity or discontinuity turn on the usefulness of the characterization of the phenomenon, not upon onto-logical assertions about nature being truly discrete or otherwise. One can always narrow the measurement grain and extent, or change focal definitions, so as to find continuity where before there had been appar-ent discontinuity. Changing the data collection protocol changes the form of the data plot, but will not resolve the question as to which type of description is more appropriate. In the final analysis, discontinuity is

FIGURE 6.9   "Removing background noise, water, so as to concentrate signal, fish." *Photograph by Paula Lerner*

a decision that hinges on considerations of the relevant constraint, not a reality independent of observation.

Conclusion

In this chapter we have looked at filters as they are employed by both the observer and the observed, material system. An observation of a material system is a second-order interaction between filters: the filters of the observation protocol as they interact with material filters in the observed. Experience can be described as the interface between our observational filters and the filters of empirical structures.

Grain and extent allow only certain signals from the world to pass into view. The signal of interest, which may be energy, material, or informational, passes through the filter. Earlier we used a fishing net to explain notions of grain, extent, and scale. At first glance the fish are filtered out, and the water passes through the filter, as if it were the product of interest, which of course it is not. After this chapter, the net can now be seen as a filter, if we think of it removing background noise, water, so as to concentrate signal, fish. With the noise removed, the signal clearly passes through the analyst's filter. In this way, data collection is a filter in the conventional use of the term.

In engineering, there is common reference not to what the filter removes but to what it allows to pass. Thus a filter that disallows low frequency signal is called a high-frequency pass filter. Conversely, a filter that removes high-frequency signal is called a low-frequency pass filter. By looking at filters in terms of what they let pass, filters provide, in addition to spatial size and characteristic frequency of their outputs, a means for ordering empirical entities hierarchically. Filters associated with higher levels of organization integrate information over a longer period of time, or wider spatial expanse, than do lower levels. Entities belonging to high levels have low-frequency pass filters. Conversely, those belonging to low levels have high-frequency pass filters.

Filters in the material system play a direct role in determining what we see. Our experience of empirical entities is restricted to material or information flowing through their filters, from their output channels. Our observation protocol does not have access to things in themselves, but only to their outputs, and we ourselves are further removed behind the filter of our observation protocol. Regular structure and hierarchical order emerge from differential rates of energy, matter, and information flow at the interface between the observer and the observed. The model we propose is an alternative to the conventional model of "objective" science done by naive realists. The model here is a vision of temporary and fluctuating states of matter and energy, experienced by a subjective observer, only after matter and energy pass from the surface of the observed through the observer's input filters.

7

Defining the Whole with Surfaces

The General Nature of Surfaces

In the last chapter we laid out the notion of filter. Filters lead very easily into another central notion in hierarchy theory, the nature of the surfaces that bound entities at a given level. In much the same way as successive filters identify successive levels, it is possible to see hierarchical structure, particularly in nested hierarchies, as a set of surfaces surrounding more local surfaces within. Much of what was expressed as the effects and properties of filters in the last chapter can be translated into the effects of surfaces. This is no accident, since filters are usually identified with surfaces, surfaces being the places where filters operate. Surfaces can be conceived as filters slowing the flow of material, energy, or information. Sometimes the delay is so long that the surface completely attenuates the flow altogether. This leads to the critical feature of surfaces, which is that they are apparent discontinuities in the observed. Surfaces are boundaries around material, energy, or informational flows.

In the present chapter we relate surfaces to levels of observation, using them as an extension of the notion of a filter, but also allowing them their own distinctive explanatory power. The nice thing about surfaces over filters is that surfaces are usually a concrete part of experience. They therefore put some of the abstractions surrounding filters in much more substantive terms. Accordingly, in this chapter, there is a series of examples that make notions of level of observation more intuitively obvious.

We will use the simple example of a glass of water to illustrate how surfaces characterize the entity they surround by allowing information to pass at characteristic rates. This, of course, puts the entity at a particular hierarchical level. We will use the example of a national border to make the same point, but also the point that useful surfaces are usually boundaries according to many criteria. Such rich surfaces are useful in developing the notion of an emergent property, which we will soon define. The formation of surfaces, along with the creation of hierarchical structure, is covered in the example of the surface between cold and warm water in lakes. The argument pivots on relative reaction rates, and this in turn leads to the separation of entities in a temporal frame, as opposed to obviously spatially characterized surfaces. Biochemical cycles, and class structures in society, are the examples here of temporally based distinctions. Relative reaction rates translate to relative strengths of interaction at different levels in a hierarchical system, and the examples used here are neurons at a lower level as opposed to the whole nervous system at a higher level. Chemical bonds are seen as higher-level as opposed to atomic bonds at a lower level. The chapter ends with a unifying and abstract conception of two-way surfaces. There we introduce a term invented by Arthur Koestler for such situations, the *holon*. But more of that, and of the other ideas above, as this chapter unfolds.

The Natural World and Natural Systems

Physical Material Systems and Levels of Observation

Surfaces allow matter, energy, or information to pass through themselves at characteristic rates. Inside a closed surface, the material flows are relatively unimpeded, giving a certain unity to what the surface contains. Consider the surface of water in a glass. The surface is the product of water molecules flowing faster and more easily inside the liquid than they do into the air. The chemical bonds in the cup hold the water together in a loose association, and the bonds, en masse, produce a high surface tension in the liquid. Notice that the passage of individual water molecules into the air is not completely prohibited, but is only slow relative to flows within the cup. In the end, if the cup is left alone long enough, the water will all evaporate. Thus the water's surface results from a delay in material transfer between the cup and the air. By slowing down the exchange, the filter sets time constraints on the entity. In this way, filters give each surface a characteristic rate of exchange, and the surface correlates with the boundary of the observed, empirical entity.

The flow of water is relatively fast, as liquids go; some liquids flow much more slowly, like molten rock in a lava flow. Glass is, technically speaking, a liquid—one whose rate of movement is particularly slow at room temperature. The rate of flow in glass is sufficiently slow to allow it to be treated as a solid. Despite its apparent solidity, the essential liquid nature of glass is apparent in the distortions one sees through old windows. The windowpane is literally flowing out of the window frame, and the distortions are the effect of ripples in that flow. Tip over a glass of water, and the water flows away, but the glass moves so slowly that flowing would seem a misnomer.

Given the slow rates of flow within glass, it is hardly surprising that the rate of exchange of glass across its own surface, into its surroundings, is much slower than the flow across the surface of water, as water evaporates. Slow as it is, glass molecules do escape into the surrounding medium. Much as there is flow within glass over years, glass molecules will also pass, over a period of long years, from the mass of glass, through its surface, into its surroundings. Biological specimens kept in glass jars undisturbed from the nineteenth century are sometimes a bit crunchy, even though they were soft animal tissue when they were put in the jar. The glass has literally been dissolved in the preservation medium and then redeposited on the specimen. The filter associated with the surface of the glass jar is, for a liquid, a very low-frequency pass filter. Translating this to levels of observation, the low-frequency behavior of glass makes it a very reasonable contextual level for water, a fast-behaving, low-level material. That is why glass is often the context for water, in the form of a drinking vessel or bottle.

Boundaries in Human Systems

The examples of water in a glass, and of glass itself, come from the realm of tangible material things. The same points can be made by looking at human social systems, so the above concrete ideas can be translated into the realm of human artifact. Much as physical objects can be ascribed to levels by virtue of their surface properties, human social systems can be placed in ordered levels by their respective surfaces.

Consider the example of jurisdictional boundaries: neighborhood, city, state, and national boundaries. Let us explore the manner in which a national boundary coincides with that of a border town. The town has

FIGURE 7.1 The Detroit waterfront seen from Canada across the Detroit River gives a spectacular view of the United States. *Photograph by T. Allen*

two types of limit; one between it and its surrounding state, and the other city limit that faces across the border to another nation.

The national boundary, as opposed to ordinary town limits, deals with the flow of information in a manner that reflects the national standing of the entity it circumscribes. Information and material that does cross the surface characterizes an entity belonging to the "nation state" level of organization. Mail is a good way to illustrate the point. Of the mail from inside the country that reaches a border town, only a small amount is destined to cross the border. Strong connections within, but weak connections across, define the surface. This is very like the considerable flow inside a body of water, and the limited flow of evaporation across the water's surface. In the case of mail, there is more to the story than just the quantity of mail from given sources, and this shows how the national border has different qualities from the character of some lesser political boundary.

Not only is there less mail crossing the border, but that which does cross characterizes an entity of national standing. Mail crossing a national border comes much more broadly from the country of origin than mail that is directed to a border town within the country in question. Proportionally, the mail crossing into Canada is better integrated across the U.S.A. than mail directed within the country to border towns like Detroit (figure 7.1). "Motown" gets most of its mail from the neighboring Upper Midwest, the state of Michigan, and the closely bordering states of Ohio and Indiana. Of course, most of the U.S. mail that does make it across to Windsor, Ontario, also comes significantly from the Upper Midwest. However, because Windsor is in Canada, within the subset of its mail that does come from the U.S., the proportions are not so heavily biased in favor of the Upper Midwestern states when compared to the overwhelming mass of mail from Michigan, Ohio, and Indiana that enters Detroit, Michigan. The absolute quantity of U.S. mail is smaller entering Windsor, but, just as important, the ratio of states represented will be more evenly distributed across the entire United States. Proportionally, a letter coming into Windsor in Canada stands a better chance of coming from a distant state, such as New Mexico, than does a piece of U.S. mail entering Detroit and staying within the national boundary. That which crosses the boundary tends to pertain to the entire large entity circumscribed by the national boundary, not to the local environs of the portal that was used to make the exit.

The International Joint Commission of the United States and Canada has a regional office in Windsor, Ontario, but it has a mail drop in Detroit as well as a Windsor mailing address. Someone from the regional office in Windsor couriers across from the mail drop most days. This saves a lot of time, since letters from Detroit sent through the normal mail route to Windsor go via Toronto, several hundred miles from either city.

The point of the example is that boundaries integrate the entities that they contain. The U.S. and Canada are separate entities, both characterized by a high degree of interaction within each entity, and less frequent, weaker interactions between entities. Much as liquid water becomes water vapor as it crosses the surface in the process of evaporation, mail leaving the U.S. has a different character from mail that stays in the country. In hierarchies, surfaces interface between parts, wholes, and the larger context. Surfaces lie between parts at the level of parts. Surfaces also lie between wholes, as a collection of parts, and the outside world. Surfaces integrate wholes, and give them the properties that characterize the level to which they belong. The contrast between the different boundaries of Detroit, one with the rest of Michigan and the other with Canada, shows the difference between surfaces that surround low-level parts, as opposed to those that surround high-level wholes.

Natural Surfaces

There are many examples of surfaces, and they can all be related to flows of information, matter, or energy. National borders are a case in point. Flows of automobiles, culture, money, mail, and people may all reach the national boundary, but most money and people, with their respective languages and values, do not cross it. The border corresponds to a change in rate of interactions for a large number of different processes. While mail was the example used immediately above, we could just as readily have used one of many other criteria to characterize the international frontier. The fact that the border coincides for many facts of human activity makes national borders a member of a special class of boundary, one we call "natural boundaries."

Natural surfaces do include man-made divisions. The border between countries is not a reflection of some law of nature. Nevertheless, it is something with very real consequences. The caprice of war and treaty may place the boundary where it is, but once it is there, all sorts of criteria order themselves around it, and reinforce one another. We use the national border example here by no accident. Because it invokes an anthropogenic boundary, we can force home the point that "natural boundary" is not an indication of something particularly ontologically real, or a product of some pantheistic, bucolic setting.

"Natural" here means the same as it does in a natural classification. A natural classification is an ordering that reflects a large number of criteria that all coincide to give the categories or classes. Similarly, natural surfaces coincide for a large number of criteria rather than indicate something ontic or in accord with nature. The filter that corresponds to the surface works on a large number of different types of signal. Natural entities have a certain robustness. Observe them according to one criterion and they can still be seen according to another. Watch the people or watch the money; both generally move more inside the border than across it. Note here that observational considerations are again central to the discussion. A natural surface is one that appears to remain in the same place, even when one changes from observing one type of signal to observing another. Since the filter at the surface works on a large number of signal types, the surface appears in the same place when observed under different criteria.

The coincidence of many criteria at a natural surface often arises from mutual reinforcement of the processes that are associated with the several criteria. In the case of a national border, there is reciprocal support in limited flow of people, giving limited cultural exchange, which supports cultural differences, which return to diminish interest in crossing the border to the alien culture. These mutual reinforcements might appear of little significance in well-established borders, such as the example we used. The U.S.-Canadian border is the longest undefended border in the world, and the mutual reinforcement of various social processes there merely maintains a firmly established status quo. United States citizens are often not aware of the remarkable unity of Canada, given the narrow southern strip of Canada that is heavily populated. Canadians regularly move back and forth great distances across their country from the Eastern Provinces to the Pacific Coast as job opportunities come and go. For all that movement close to the border with the United States, there is relatively little movement into the neighbor to the south. When Canadians go abroad, they often hop right over the United States, and vacation in winter in the Caribbean.

When borders move, the importance of mutual reinforcement becomes clear. As the peoples of Bosnia separate according to religion, those lines of demarcation will generate many other criteria for separation along those same geographic boundaries. Conversely, the tension that exists in the Baltic nations over minority Russians who moved to the Baltic inside the former Soviet Union is the effect of suspending multiple criteria for separation wholesale. The tragedy of the cultural destruction of Tibet is another example of deliberate suspension of as many criteria as possible for separation from the rest of China. It is in recognition of the many different types of disruption that occur when borders move that one of the guiding principles of the United Nations is the inviolability of national borders. International corporations begin to emerge as a major force for a change of world order, as they insert themselves at a level of organization that is yet higher than that of nation states. By manufacturing the same product in several different countries, international corporations move national boundaries from above their sphere of influence to within their purview. They can hold off national government threats of regulation with the capacity to switch production to a factory in another nation. Corporations do not need to use threats themselves, for all governments can make the calculations.

Emergent Properties

An important issue in systems analysis, related to multiple criteria for surfaces, is the matter of emergent properties. In a trivial sense, emergent properties are properties of an entity that one did not expect. More significant are emergent properties that come into play only once one is concerned with a wide enough scope. For example, biological phenomena cannot occur in a universe of study so small that it admits only simple inorganic chemistry. It is a mistake to imbue the notion of emergent property with more significance than it deserves. Emergent properties need not be something mystical or mysterious. If systems zealots yield to that temptation, systems theory is laid open to criticism from obligate reductionists that the paradigm lacks rigor. Emergent properties are part of commonplace experience. John Searle uses the example of water. It is not possible to find the emergent property of wetness at the level of individual water molecules. As Searle puts it, one cannot go in and carefully pull out a water molecule and say, "Ah! This is a wet one." It all makes intuitive common sense and is part of mundane experience, in that water vapor consists of the same stuff as liquid water but is not wet, because the water happens not to be liquid at that time. Searle cuts the Gordian knot of the mind-body problem in the same way. Mind is an emergent property of brain tissue, although one cannot go in and find a neuron firing, and identify it as a thought.

In the context of natural surfaces, the multiple criteria that align with such surfaces are emergent properties. Find a structure, then change the criterion for observation and see the structure persist. David Bohm has said that the very purpose of science is to find that which is robust to transformation. The transformation is a change in focus from one process that is constrained by the surface to another. Natural boundaries are robust because they persist across different observation regimes. That persistence depends on several observational criteria all sharing the same natural surface.

Science is interested in being able to generalize from a limited set of observations. Thus observation of a natural surface on one criterion allows generalization to the other criteria that depend on the other types of signal. That generalization is a prediction of "If I see such-and-such a signal in a given place, I can predict that these other signals will coincide." The ultimate generalization is prediction into the future. In order to make predictions, it is necessary that the entities whose behavior is to be predicted should be stable for the period of the prediction. In the upshot, natural surfaces tend to be stable by virtue of mutual reinforce ment of the many processes bound by the surface.

The Dynamics of Surfaces

The Formation of Surfaces

Surfaces will spontaneously form when there is a significant gradient in concentrations of information, energy, or matter. Surfaces amount to local places across which there are significant differences. In the absence of a surface, any differences appear as gradual changes over wide spans. Across these wide spans there are continuous gradients. In wide regions of large but continuous difference, the local flow often organizes itself into cells with surfaces in between them. The difference that occurred, spread out along the entire gradient, becomes localized and concentrated in only one small stretch of the gradient. That layer where the difference is localized is the surface itself. On either side of this new surface, the gradient flattens out. All the difference of the long gradient is concentrated at the surface, leaving homogeneous regions on either side. Nature appears to abhor a gradient and often will do what it must to

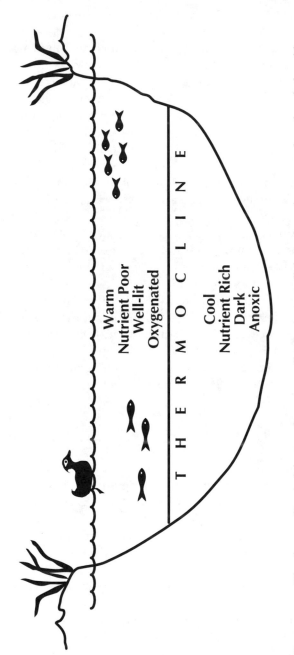

FIGURE 7.2 The thermocline in lakes produces two cells of water, one light and nutrient-poor, the other dark, cold, and nutrient-rich.

localize gradients at surfaces. The steeper the gradient, the more likely it is that a surface will emerge.

A simple example of a surface that emerges when a gradient becomes steep enough occurs every summer in temperate lakes. This surface isolates an upper layer of warm water from a lower layer of cold water that occurs in lakes in the summer (figure 7.2). One can encounter it when treading water. Your torso is in water of a comfortable temperature, but your feet kick down into the cold layer below. This happens only in the summer, for in the springtime temperatures in the lake grade continuously from top to bottom. In spring any differences in temperature of the entire column are small because the water is mixed throughout the lake by March winds.

The surface that divides a warm upper layer from cold water below forms suddenly in early summer. As the winds die down, when March goes out like a lamb, the tendency for warm water to rise to the top increases. Furthermore, water at the top, warmed by the sun, tends to stay on top, so the temperature gradient top to bottom is continuous but spans a wider difference in temperature. Soon local eddies occur so that cold water comes up against distinctly warm water in the lake, to make the beginnings of a surface. Because the temperature gradient is steep, all of a sudden, one of these chance occurrences propagates itself across the entire lake. In just hours, a warm water layer comes to overlay, and is disjunct from, a cold water layer underneath. The surface between them is called the thermocline. The thermocline is a natural surface, because there is little exchange of water, nutrients, oxygen, or biota across the surface. When your feet experience especially cold temperatures in treading water, it is because they have kicked through the thermocline.

Surfaces in Time

In the thermocline example, the surface is spatially identifiable, but it arises because of relative rates of flow of water. The thermocline is a local place in space displaying local rates of flow. The relatively fast flows inside the upper warm layer contrasts with the slow flow of warm water down through the thermocline. Thus the thermocline is a place in space, but it depends on relative flow rates, which are distinctly temporal.

An important aspect of surfaces is the way that there is strong interaction inside a surface and weak, sluggish interaction across. For example, it is that property of fast reaction within a surface but slow reaction across it that is significant about a biological cell membrane. Cells have integrity because the parts interact often and quickly inside the membrane, while the environment is excluded and can exert an influence only over an extended time. It emerges that so long as the relative flow rates are retained, then the spatial integrity of surfaces becomes incidental. Spatial integrity can be cast aside, and so long as there are very different relative rates of exchange, there is still a functional boundary.

While cell membranes can be literally seen as things in space, some other aspects of biological systems at a cellular level are not localized in particular places. Biochemical cycles are a reliable part of cell functioning, but they occur somewhat diffusely inside cells. Some aspects of biochemistry do occur on particular membranes, but a lot happens out in the body of the cytoplasm. In the cell, biochemical cycles are entirely coherent, but that coherence relies on relative reaction rates rather than being spatially localized. The biochemical cycle is bounded not in space but by relative reaction rates. Reaction rates are strong and fast within the cycle, but weak and slow outside.

Biochemistry is different from dead organic chemistry because biochemical interactions are ministered by enzymes. Enzymes are proteins that do not change as they participate in reactions, but their presence increases the rate of the reaction by an order of magnitude or so. Enzymes are the means whereby biochemical reactions are completed before ordinary organic chemical reactions can have an effect. In this way, enzymes separate the living from the dead not by spatially interposing themselves as a physical surface but by having the same temporal properties as a physical surface. Enzymes facilitate strong connections inside a biochemical pathway, and leave normal organic chemistry as a torpid background in which the biochemistry takes place. As a result, biochemical pathways cannot be found as tangible things in places, but they still have integrity by virtue of vigorous interaction between parts, and weak interaction with all else. The material of biochemical pathways are mixed together with a matrix of other chemicals in the cell's cytoplasm, but the pathways remain functionally separate, just like the warm water circulating above the thermocline is separate from the cold water below (figure 7.3).

The same can be said of social surfaces that bound such intangibles as social class, nationality, culture, and religion. True, there are physically identifiable structures associated with these social entities, like gentlemen's clubs, servants' quarters below stairs, separate lines for aliens at

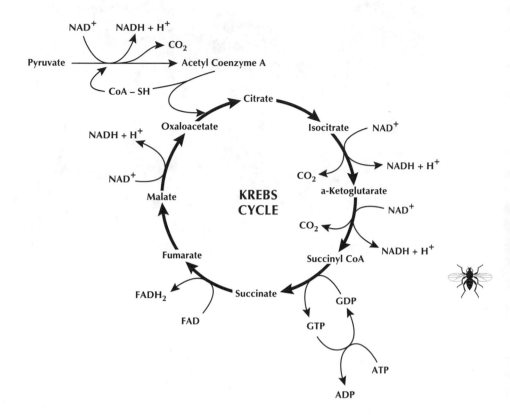

FIGURE 7.3   "Biochemical pathways are mixed in with a matrix of other chemicals in the cell's cytoplasm, but the pathways remain functionally separate."

ports of entry, and cathedrals for each separate denomination. These social surfaces retain integrity by more frequent interaction within the group than by any physically bounded place. True, social classes and religions do lose members because of contact with other groups, but that is because social interactions are not closed. Social interactions are, however, stronger within groups than between groups. The point is that all intangible but nevertheless distinctive entities are bounded by surfaces characterized by relative rates of exchange of important material, energy, or information. The boundaries need not be identifiable as a tangible boundary in space, but can function entirely on different rates of

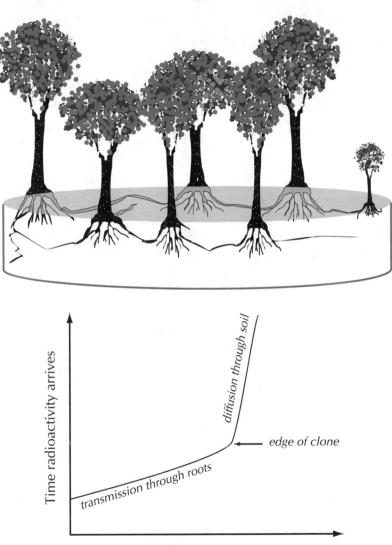

FIGURE 7.4 A clone of trees is strongly connected by the roots, and only weakly connected to the outside ecosystem. A graph of time since introduction of a signal and the spatial displacement from the point of injection at which the signal is detected is more or less a straight line, until the edge of the clone. Then the line turns upward, indicating that the signal transmitter has changed to new slower process, plain unchanneled diffusion.

exchange within the abstractly bound entity compared to the rates of exchange with the surroundings of the abstract entity.

Surfaces and Levels of Organization

Flow Rates at Surfaces and in Communication Channels

We have seen in earlier chapters that the level of organization at which an entity resides comes from either the general characteristic behavior of empirical entities that occupy the level, or from the responsiveness of the filter that is located at the surface. The level is not the surface itself, but is a generalization of the properties of the surface. That is a different matter. For example, a particular whole human being is not a level in itself. The level could be the level "human," or the level "organism" to which a large number of other organisms belong. The properties shared by the group that prescribe the level are more general than the individ- ual examples themselves. The level is a different logical type from the things that populate it. Each example is no more the level than is a par- ticular book the concept of "book." Therefore, the general properties of surfaces as they define levels are more inclusive than the whole entities one perceives.

Thus surfaces are places in the world where there are changes in the amount of interaction between empirical entities. Another example will help here. Poplars reproduce by sending out roots underground that sprout new and apparently separate trees. A group of such vegetatively spreading trees is called a clone, but where is the surface of that clone (figure 7.4)?

Tritium is radioactive double heavy hydrogen. If one were to inject tritiated water into a member at the center of the clone, then radioac- tivity would spread quickly through the underground stems to other members. First the radioactivity would spread to the injection site's immediate neighbors, and then to more distant trees. The farther away from the point of injection, the longer the tritium takes to arrive. The delay in its transmission could be seen on a graph of distance from the point of injection, against time taken for arrival of radioactivity. The graph would be more or less a straight, gradually sloping line, but only up to a certain point. At points beyond the edge of the clone, the radio- activity would be considerably delayed, because organic connection stops at the surface of the clone, separating the clone from the grassland beyond. At the surface of the clone, the rate of transmission of radioactivity changes dramatically. On the graph of time against distance, there is a sharp steepening of the plotted line, indicating that a surface has been reached.

Interaction inside the clone is strong, but interaction across the surface of the clone is weak, reflecting relative process closure. Processes within boundaries are strong interactions that have a degree of closure imposed on them by the surface of the clone. Weak interactions across surfaces belong to processes not included in the closure. In the clone of poplars, the weaker process across the surface is one of leakage from the roots to the soil, which is much slower than the exchange of water, salts, and food through the active transport system in the connecting roots.

At this point let us introduce the concept of a *communication channel*. Communication channels are the opposite of surfaces. They are defined by how they let information pass. Surfaces and communication channels represent different sides of filters, because filters may be seen in terms of either what they block or what they let pass. Whereas movement across a surface involves a great change in rates of interaction, movement down the great length of a communication channel involves no change in rates of interaction whatsoever. When the signal of a voice has traveled any distance down the cord of the handset of a telephone, the signal can then travel across great distances between continents with no delay. Communication channels are also part of the poplar clone example.

In the clone of poplars the connecting roots are communication channels between the entities at the lower level, the individual trees in the clone. Once the tritiated water is injected into any plant, the transport system moves it to other parts of the clone relatively fast. At each level, entities have a set of communication channels between parts. The whole, when it functions as a part of a yet higher-level entity, has access to another set of communication channels that bind the whole to other entities at its own level. These communication channels are those inside a still higher-level entity, of which the whole is itself a part.

Surfaces are local, and communication channels are global. Passage through a surface changes temporal characteristics of signals. Communication channels, on the other hand, preserve temporal relationships of signal streams passing down them. Communication channels to the outside are the means whereby an entity becomes part of a larger system, whereas surfaces define the entity at its own level.

Surfaces and Strength of Connections

The discussion above that centers on rates of communication can be readily translated into strengths of interaction. For rate of reaction, read strength of connection. The strength of connections depends on the level from which they are observed. Entities populating lower levels show strong connections within their surfaces. These connections pertain to processes whose interaction form the low-level entities in question, which are here considered as parts. Relative to the strong connections that give the parts their separate identities, the connections that bind the parts together to make the whole are weak connections. However, at the level of the whole itself, the connections between parts give the whole its identity. Going further up the hierarchy, the whole does have its own weak connections out through its surface. These are, of course, the strong connections inside an entity at a still higher level. And so on. Strong connections within give integrity, and weak connections across bind upper-level wholes. Thus at all levels, when one looks at a situation from the level in question, there are strong connections inside a surface, and weak connections across a surface. Surfaces separate entities at the level where the strong connections function.

Seen from outside the surface, the weak connections across the surface give the entity its characteristic properties. The weak connections not only link the entity to others that exist at its own level (other empirical entities), but also link the entity to the observer in an observation. An example of strong and weak connection will help here. The long axis of a neuron cell in the brain is called the axon. Once a neuron starts firing from one end, the signal passes down the full length of the axon without being diminished. However, when the signal reaches the other end of the axon, it must pass as a chemical signal across a gap, called the synapse, to receptors of other cells, if the signal is to go farther (figure 7.5). Often the chemical signal can get into the gap, but if it is not sufficiently concentrated it will fail to activate enough adjacent neurons to cause a further propagation of the signal.

Relative to the propagation of electric current down the length of an

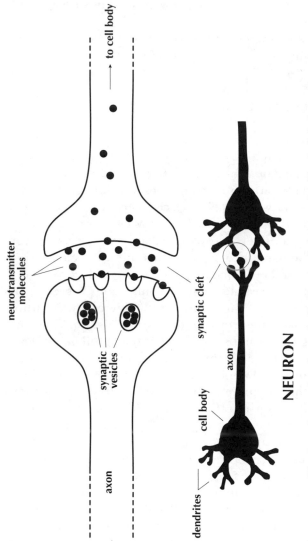

to cell body

axon

neurotransmitter
molecules

synaptic
vesicles

synaptic cleft

axon

dendrites

cell body

NEURON

FIGURE 7.5 The impulse moving down the long axis of a neuron is electrical. It depends on a feedback of ionization. At the end of a neuron, the only way a signal can pass from one neuron to the next is through chemical signals. The change from ionized to chemical signal is a change in levels.

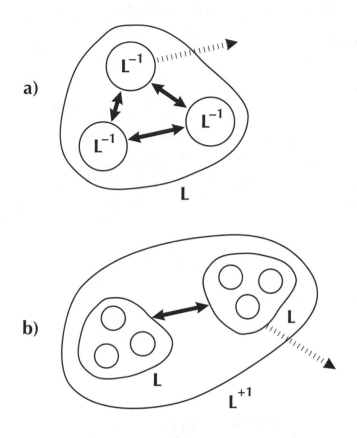

FIGURE 7.6 In a) the strong connections between members of the lower
level are clearly shown, with the weak connection to an equivalent higher-level
aggregate. In b) the weak connections to aggregates at level L become the
strong connections inside the entity L + 1.

axon, it is only when sufficient activity is present that an impulse ampli-
fies and activates higher levels in the nervous system. At the level of the
individual neuron, the chemical links between cells is weak. However,
seen from the higher level of the entire nervous system, these chemical
connections are strong and provide the integrity of the system. From the
level of the entire nervous system, passage of electrical signal down individual axons is taken for granted. The strong interactions of the upper

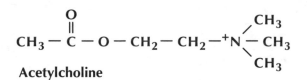

Acetylcholine

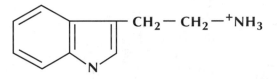

Dopamine

$$\text{CH}_2 - \text{CH}_2 - {}^+\text{NH}_3$$

Serotonin

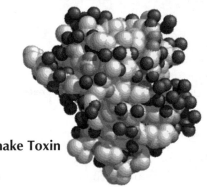

Sea Snake Toxin

level, the brain, are in fact the weak interactions that pass across inter-neuronal surfaces of the next level down. Neuron endings are spatially separated at the lower level, causing weak neuron-neuron interactions, but they simultaneously provide the strong connections that unify the upper level whole.

There are two types of separation in the preceding discussion. First, there is a separation between individuals at the lower level, which we shall call horizontal separation across the level in question. Second, there is a separation between the lower and upper level, that we shall call vertical separation of levels in the hierarchy. Surfaces horizontally separate entities at a given level from each other, but they also separate levels when the surfaces separate each lower-level entity from its respective general context (figure 7.6). In the above example, horizontal separation would be between individual neurons at their own lower level, or between the brain and the skull at their higher level. Vertical separation between levels occurs as each neuron is separated by the gap across the synapse from the rest of the brain. The differences between levels are differences in frequency characteristics; upper-level communication behaves with a longer periodicity than lower-level communication. Not only is the passage of signal between neurons slower than down an individual neuron's axon, but it is also completed less frequently. Not every release of neurotransmitter causes the firing of the next neuron down the chain (figure 7.7). Moving up one level further, no activation spreads from the brain to adjacent parts of the skull.

Observations are made from the outside, and it is the surface that causes the entity to display this as opposed to that behavioral frequency. All neurons transmit signal at close to the same rate; all connections between neurons transmit signal at about the same rate, but it is at a slower rate than transmitting signal down the length of an individual axon. Behavior at the level within neurons has a different characteristic frequency from the level of communication between cells within the nervous system at large. Lags in transmission across surfaces characterize the entities at a given level, and it is cumulative signal frequency that defines

FIGURE 7.7 The standard neurostimulators. They have different roles to play in a normal nervous system. The bottom panel shows the three-dimensional molecular form of a sea snake neurotoxin; it is a mimic of acetylcholine. It saturates the receptor sites, outcompeting the molecule it mimics. The neurons cannot stimulate one another properly.

FIGURE 7.8 "Janus is portrayed as having two faces, one looking in and the other looking out."
Photograph by Paula Lerner

entities at a given level, and it is cumulative signal frequency that defines one level as separate from an adjacent level.

Bond Strength and Integrity

Let us now translate the above discussion of strength of connection into one that deals with the relative integrity of entities at their respective levels. Lower levels of organization have higher bond strength and integrity relative to upper levels of organization. Take, for example, molecules as a level of organization that is built from chemical bonds. The individual atoms in a molecule represent a low level of organization. Accordingly, the atom is more integrated than the molecule, and so it has more energy associated with its internal structure. When bonds are broken, there is commonly a release of energy that corresponds to the energy that was embodied in the strength of the connection. Break a high-energy chemical bond, and a release of energy can be measured. It is exactly on that principle that plants work when they trap energy in photosynthesis and release it in respiration. It takes sunlight energy to make the bonds of sugar molecules from carbon dioxide and water. Respiration breaks the sugar down to carbon dioxide and water again, and the energy that is released is then used for repairing and building the plant.

The reason that an atomic bomb or hydrogen bomb releases so much explosive energy from such small quantities of matter is that bonds within atoms are manipulated. Compare the energy released in an explosion of a few pounds of TNT with the explosive capacity of the actual core of an atom bomb. When nuclear weapons are described as being so many megatons, it is an indication of the enormous equivalent mass of TNT that would be required to make the same explosion. The difference in mass is an indication of atoms operating at a lower level of organization as opposed to the molecules oxidized in a chemical explosion.

To take another example, the integrity of a conversation comes from the connections between minds that comes from exchanges between vocal chords, ears, and cognition. The connections are clear, but there is enough slack in the lines for talking at cross-purposes, or just not hearing all the words. However, the integrity of the bodies of two individuals holding the conversation is much stronger. The individual people are the parts belonging to a level below that of the conversation. Accordingly, instead of the tenuous connection through sound waves, the parts of each individual are tightly connected through muscle, sinew, and blood circulation. Bond strength, therefore, provides a means for ordering and comparing levels. Entities at higher levels of organization have less bond strength than the parts that comprise wholes at lower levels.

The Holon

Parts are set in the context of the whole to which they contribute, and lower levels of organization are set in the context of higher levels of organization. The notion of parts versus wholes generally invokes a nested hierarchical system, the properties of which were discussed in chapter 5. The limitation coming from parts below is subtly different from constraint coming from levels above. The parts indicate what is possible, given the stuff of which the system is made. Constraints, on the other hand, impose order on the possibilities that come from below. These limitations define what the structure of the whole allows the parts to do. It is this distinction that makes biology more than complicated physics and chemistry. There are limits imposed by the biology of an organism that allow masses of carbon and other essential elements to do only certain things. For example, it is possible for carbon to form into a very stable tetrahedral structure, a diamond, but that is never part of biological process. The sorts of pressures and temperatures needed to make a diamond are not allowed inside the constraints of biology. The behavior of the whole cannot be something of which the parts are incapable. If the relationship is one of part-whole, the behavior exhibited at upper levels is limited by what is possible at lower levels.

Capturing the set of relationships between the whole, its parts below and context above, is one of the goals of hierarchical analysis. Arthur Koestler, in *The Ghost in the Machine,* saw entities as two-way windows between their parts and the larger whole of which the entity is itself a part. Entities may be seen as both parts and quasi-autonomous wholes, both at the same time. Koestler even coined a new term for entities defined in that general way: *holon.* Poetically, he called them Janus-faced holons, after the Roman god of portals, windows, and doors. Janus is portrayed as having two faces, one looking in and the other looking out (figure 7.8). Entities defined in this way are a skin, an interface.

Koestler's description of entities as holons is wonderfully economical in capturing the relations between the observed and the observer. The rest of the universe experiences the parts of the holon only through the way that

they work together to make the whole. From outside the holon, the parts may be seen as integrated to make the whole. However, inside the whole, the parts experience outside influences only as they are modified by entering the whole. For the parts to experience the world outside, signals must penetrate the surface of the holon. In passing the surface, moving either inward or outward, signal is commonly filtered and modified.

An example of how signal is modified by a surface is the balance of sunlight as it enters the ocean. Ocean water appears blue because the red end of the spectrum is absorbed. Only blue light travels freely into water to be reflected out again. Sound quality is likewise changed as it enters a building. An argument in the next room is muffled because the higher-pitched sounds are absorbed by the walls, while only the lower-frequency parts of the sound pass through. At surfaces input channels can even completely change the medium of the message into some different stream of energy or matter. The eardrum vibrates to the stimulus of sound, but it is the electrical signal of the auditory nerve that actually enters the body to be perceived as sound. Similarly, light stimulates rod and cone cells in the retina, which in turn transform electromagnetic energy into an ionic impulse wave that reaches the optic nerve and travels to the brain. Telephone systems likewise convert sound vibrations to electrical or light impulses. In all these examples external signal is modified at a surface before it is experienced by the parts inside. The nature of the experience within the whole is, however, significantly different from how the whole is experienced from without. Having a sensation of hearing or seeing is significantly different from wondering whether another person can hear or see. The possibilities and constraints that allow sensory experience within a perceiver are significantly different from the possibilities and limitations on our knowledge of the sensory experiences of others.

Conclusion

Surfaces are the places in the world where there are changes in interaction rate and strength. In technical terms, they are the steep parts of an interaction density gradient. It is at the surfaces of structures that we find

---

FIGURE 7.9 "It is at the surfaces of structures that we find the wrinkles in time and space that give the world texture." *Photograph by Paula Lerner*

the wrinkles in time and space that give the world texture, and allow us to conceive and perceive it as hierarchical (figure 7.9). Surfaces correspond to disjunctions in the material world where an observer specifies this as opposed to that continuous gradient of change as being steep enough to warrant it being called a boundary. While we need not suppose that the material world is fundamentally discontinuous, it does appear that some continuities involve steeper gradients than others.

The strong connections inside a surface are important in defining an entity as a set of parts connected in a certain way. By contrast, the weak connections that do pass through the surface give the characteristics of the whole that is bounded by the surface. While surfaces have a lot to do with how things appear, like complexity and stability, they are a product of observation. The entity in question might at first appear through weak signals passing outward from the observed, but analysis and definition comes from decisions as to which strong connections within make up the whole in the observer's model.

Surfaces introduce a set of dynamical considerations into matters of structural boundaries. The earlier chapters emphasized how structure is a matter of observation protocol and decisions. Observation is a matter of setting up filters through which the world is perceived, and it is the relationship between the observer's and the observed's filters that determines what we see. Notions of filters and surfaces are central in bringing tractable order out of a continuous stream of experience.

8

Evolution and Revolution in Creating Complex Structures

We have been at pains to emphasize that hierarchical structure may not be a manifestation of a material hierarchy, but that it emerges as part of the process of observation. However, that line of argument begs the question of why it is that we experience hierarchical structure so often. Herbert Simon has identified that if there are complex structures in the world that are not hierarchical, we would have no way to know them. This chapter is concerned with the processes that are responsible for the regular appearance of hierarchical systems.

There is a parable in hierarchy theory emanating from Herbert Simon that goes a long way to explaining why the external world is full of systems that lend themselves to hierarchical explanation. It is a tale of two watchmakers. One watchmaker is hierarchical in his method and the other is not. Both watches consist of a thousand pieces, but the nonhierarchical timepiece must be fully constructed before it is stable. If the watchmaker is forced to put the watch down before it is completed, then it falls apart, and he must start again when he returns. By contrast, the construction of the hierarchical watch is as a set of small, stable subunits. When the first stable unit of ten pieces is made, the hierarchical watchmaker puts it aside and starts work on the second ten-piece unit. The first phase is complete when he has one hundred units of ten pieces each. The second operation is to put together ten of the ten-piece units to make a one-hundred-piece unit. Putting the first one-hundred-piece unit aside, the watchmaker constructs the second, eventually creating ten

FIGURE 8.1 "The emergence of new species happens when surfaces such as mountain ranges or oceans cut a burgeoning of individuals away from the main group." *Photograph by Paula Lerner*

such one-hundred-piece units. The final phase is to put the ten large units together to complete the full one-thousand-piece watch.

Both watches are excellent, and word spreads. Customers keep interrupting the men, so that they keep losing work completed. If the nonhierarchical watchmaker is interrupted, then he loses all work on that watch. The hierarchical watchmaker, on the other hand, loses only the time invested in putting at most nine units together. The nonhierarchical artisan will not be able to make another watch, but his competitor is hampered only marginally.

For evolution the message is this: any system that loses all ground gained every time there is a setback does not have time to evolve. Therefore all highly evolved systems, including living systems, must of necessity be hierarchical. They must be composed of stable subunits. If, at the failure to achieve multicellularity, life must return to simple organic molecules, then life as we know it would be impossible. However, that is not the way life works. Failure of any upper level means only retreat to the next lower level, populated by stable unicellular entities. A series of lower levels persists in primitive and advanced life forms. We ourselves are made of cells with biochemical pathways that represent adherence to some of the most primitive ways of biological functioning. In nonliving systems like large atoms, if failure to achieve some stable atomic form led back to isolated electrons, protons, and neutrons, then none of the large atoms could exist.

The surfaces of the previous chapter are crucial here, for they are what keep the lower-level subunits stable. By obstructing the passage of signal, biological surfaces allow some organisms to survive while others are selected out. It is no accident that speciation, the emergence of new species, happens when surfaces such as mountain ranges or oceans cut a burgeoning collection of individuals away from the main group. Dissection by surfaces allows change. Without isolation, the mass remains stuck at some average condition.

Incorporation of Disturbance

Disturbance has a crucial role to play in the evolution of complex systems. Disturbances are forces that come from outside the thing disturbed which assault the current configuration of processes. After the disturbance the recipient may well be modified. Should the disturbance return, it will address this modified entity. For example, fire is a disturbance, and a plant community is changed by being burned. When fire returns, it finds a modified vegetation consisting of plants that either survived the previous fire or moved quickly into the open space left by the flames. By continuing this process of modification, successive fires deepen the accommodation to the disturbance, until fire ceases to be a disturbance. Note that the destructive force of the flames has not changed, but rather the nature of the target has been altered. Only species that resist fire or return quickly after it are left. Fire ceases to be a disturbance because it has become incorporated into the system as a normal, working component. Essentially, a new, more inclusive, system has evolved. The new system has evolved to an emergent, higher level of organization.

Although it is called *The Origin of Species*, Darwin's book does not actually address the question of speciation. His great work focused on competition as it makes small changes within species. Competition is a very local consideration that usually has a conservative effect of removing aberrant individuals that would be the foundation for speciation. Competition is not a process promoting radical change, and it is often a force that stands against change. By contrast, incorporation of disturbance is not a conservative process, for it produces change whenever it happens. Therefore incorporation of disruptive processes in emergent hierarchical levels could well be a more significant means of evolution than competition for resources, the cornerstone of the traditional Darwinian model. Since incorporation of disturbance produces radical changes it is a process that should apply to creating larger-scale distinctions, such as new species.

It is no accident that Darwin's world was full of small Victorian businesses competing for limited resources. Darwinian competition is precisely for limited resources on a small scale. The business climate has changed over the last 150 years. We now live in postcompetitive corporate capitalism, where corporate takeover is normal, and companies that share a market will share a design, as do Apple Computer and IBM at this moment. When young computer hackers crack the code of the telephone company, they are not often prosecuted, but are instead given a job. Similar to Darwin's small businesses, today's corner groceries are consumed by warehouse grocery chains. Even the Queen's own grocer, Fortnum and Mason, has been forced to move away from individual orders, toward prepackaging as never before. Social circumstances often have an effect on the choice of scientific explanation outside sociology.

As a result, models of incorporation are beginning to appear in the evolution literature.

By encapsulating previously disruptive processes, the system develops the capacity to wait for problems to go away. The burned forest is replaced with species that require periodic burning for normal functioning. By incorporation of disturbance the system becomes larger and exhibits slower behavior. A case in point is the response of the United States to the Arab nations' oil embargo in the mid-1970s. A reduction of one-half of one percent in the oil supply caused a major disruption, with lines of hostile motorists at fueling stations. Getting gas became a way of life. Two significant changes have occurred since then. First, a surplus of oil has been engineered, such that a pinch in the supply line means only that reserves that lie secure in the Western industrialized nations are drawn down. The response time of the United States has been lengthened. Second, the OPEC nations have been incorporated into the world economy such that they cannot wait for the oil-consuming nations to succumb to their demands. The oil producers now need to sell oil regularly to pay for imports, all too often armaments, from the oil-consuming nations. OPEC itself is another example of incorporation; if you cannot beat the West, then the only thing to do is join it, and make an international entity at a higher level of organization. In all these cases disturbance is transformed into mutual dependence under the rubric of a larger unifying relation. The new system no longer reacts to acute shortages, because it is now a more inclusive, slower-moving context.

Changes in Level

When a system becomes unstable, upper-level constraints cannot maintain the system's current configuration. There are two possible outcomes: either the system collapses to a diffuse, low level of organization; or, alternatively, a new set of upper-level constraints emerge and the system moves to a higher level of organization. In psychology, the first outcome is pathology, as the human psyche collapses and is more vulnerable to external pressure. The second is growth and learning. In this case, learning can produce a more mature condition that is more robust, existing at a higher level. In both cases the material components of which the system was made persist through the collapse. What disappears are the relationships that held the material in some special configuration.

An example of a collapse to a lower level of organization is the death

FIGURE 8.2 "By the standards of dead organic chemistry, life is a blur of action."

Photograph by Paula Lerner

of an organism. By the standards of dead organic chemistry, life is a blur of action (figure 8.2). When organisms die, their chemical constitution barely changes. The critical shifts are in the organization of chemical material. Once enzymes no longer speed chemical reactions, then "ashes to ashes and dust to dust" involves nothing more than sinking into a sea of commensurate reaction rates. A dead body is not reactive. Therefore, even if it is made of the same stuff, it is structurally a new system because of its new internal relational configurations. Its new state involves much less organization. Another example of a collapse to a lower level of organization is the fall of Rome. The collapse of the Roman Empire did not change the populace of Italy more than other periods of strife that the Empire had formerly survived. The same people lived in Italy, so why was it not still the Roman Empire? The people may have been the same, but their relationship to one another on a macroscale was different. With a changed context, the behavior of the people, but not the people themselves, changed.

When a system becomes unstable, often it is the normal functioning of the system that tears it apart. Positive feedbacks are unstable, in that signals introduced into the system are amplified (figure 8.3). Any change of state causes a further change in the same direction. Pushed into an unstable configuration, positive feedbacks amplify without restraint from normal system controls. Unstable feedbacks in one part of the system set others in motion elsewhere. At the critical moment of collapse the systems usually exhibit vigorous, high-frequency behavior. This behavior can be directly attributed to the way that the positive feedbacks race unrestrained by the constraints which formerly held sway.

Racing positive feedbacks are a naked expression of dynamics. Constraints impose limits on how system parts respond to external stimuli. Stripped of all its former constraints, the system loses almost all capacity to incorporate outside influences. Any input signal can be retrieved in the output of the collapsing system, and any outside influence, even minor fluctuations, affect the system. Often the system continues its collapse to some diffuse, lower level of organization, as in the death of an organism or the collapse of a political system. In psychology, pathological depression in positive feedback reduces the afflicted person to a much lower energy state. The positive feedbacks consume biological, monetary, or emotional capital, and, as long as the pathological configuration persists, the system never regains its capacity to withstand external influences.

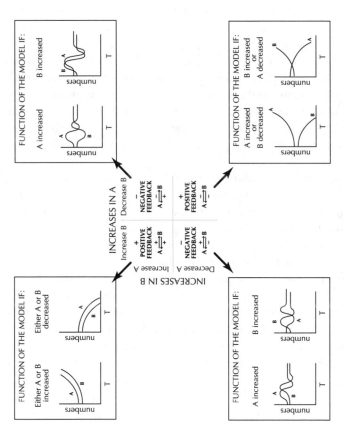

FIGURE 8.3 Feedback uses plus and minus in several ways, invoking various logical types, noted here from lowest to highest. i) The original system stimulus on one part is to push it up (plus) or down (minus). ii) A plus on an arrow means "do the same as," and a minus means "do the opposite of." iii) If a signal returns to amplify the original change, it is a positive feedback. If the signal is self-correcting, it is a negative feedback. iv) A good feedback is plus, whereas a bad feedback is minus. The scheme above is for two-part systems, but the indications are general. The chart shows all contingencies graphed in the corners.

The same initial period of excitation precedes a system collapsing to a higher level of organization. Ilya Prigogine, working as a chemist, has identified that complex systems contain a series of past instabilities. The containment of past instabilities is the very means whereby the system became complex. Lower-level dynamics that went unstable become held in a state of continuing flux in a new, larger system. The lower-level dynamics are trapped in the moment of collapse by new upper-level constraints. In a biochemical example, left to their own devices, enzymes speed reaction rates so that an equilibrium balance is quickly achieved. However, in fully functional cell biochemistry, the next enzyme in the pathway uses the end product of the first enzyme for its own reaction. Therefore the first enzyme keeps losing its end product, and so its reaction cannot achieve equilibrium. In the context of the larger pathway, single enzyme reactions are held by their context in a continuous free fall, never reaching a balance. The higher-level system has incorporated the instability as a functional component.

When collapsing to a higher level of organization, a system stabilizes around a new set of constraints that give it more organization, and a greater capacity to ride out external influences. In the process of collapse, positive feedbacks become excited to such a degree that they encounter a new set of limitations, which act as new constraints or organizing factors. This is a general description of incorporation, where the disturbing positive feedbacks become a working part of the dynamics inside the new configuration. In contrast to positive feedbacks, negative feedbacks are self-correcting (see figure 8.3). Any signal introduced is countered, so values return to their initial levels. In the system collapsing to higher levels, the new constraints are negative feedbacks that restrain the collapsing old system, using a new set of principles.

Social systems give many examples of movement to higher levels of organization. Agriculture is a case in point. Just before agriculture, the archeological record shows unusual and rapid change, with few of the constants from earlier culture persisting. That is what we should expect of a system going unstable. The collapse to agriculture was already happening, but the new constraints that hold up agricultural structures had not yet been encountered. All sorts of positive feedbacks were in full spate, two of which were population growth and expansion of trade. Vigorous trade became focused on specific sites, which became urban centers. Large centers exhibit low-frequency behavior of permanent food storage. Planting from the seed bins led to selection through

generations of plants that became more easily gatherable. The end product was new varieties, namely crop plants that had incorporated human husbandry into their biology.

Agriculture is an intrinsically unstable activity, at the mercy of the weather. Only when there was the constancy of large settlement could agriculture persist as a human activity. The trade that tore down the diffuse hunter-gatherer system itself became a central institution of the new order. With large-scale permanent settlement, the new system could survive while it waited for seven lean biblical years to pass. Note, however, that the activity inside the system is very high-frequency; marketplaces are busy.

The constancy of behavior of higher-level systems often involves more vigorous high-frequency behavior internally, but it is the constancy and intransigence of the whole that determines the level of organization. New York City is a structure belonging to a high level of organization. Note that it requires much fast electronic communication to hold together the parts of a large metropolitan system. Only after many years of neglect is the infrastructure of New York finally beginning to give way. Through the 1980s, rapid internal communication allowed the whole to deploy its large resource base to make up for a decade of neglect of cities by the Federal government.

In summary, when systems collapse to higher levels of organization, the new constraints increase the capacity of the system to wait out the effects of external influences. Whatever was the disturbance that caused the original collapse, it is now impotent in the face of the new constraints. It has done its worst, and the new constraints have met and contained that influence. The system has achieved a higher level of organization because it has the capacity to stand firm, whereas before it would yield. However, as the new upper-level entity stands firm, the internal dynamics race as if in a perpetual free fall, never allowed to come to the former equilibrium (figure 8.5).

Structure versus Behavior

The origins of agriculture example above is best described as a change in structure. It is not so much a matter of humans changing their behavior

FIGURE 8.4 Theories suggest that ancient cities, like this one, are a necessary precursor to agriculture. *Photograph by Paula Lerner*

Static Equilibrium

Balanced Forces

FIGURE 8.5. A) At equilibrium, there are static forces. b) In homeostasis, there is a mass balance of forces. Two opposing processes gives stasis. c) In far-from-equilibrium systems, the system reorganizes structurally to accommodate the increased energy dissipation. The whirlpool dissipates energy faster, and uses that energy to set up and maintain the structure that allows the efficient energy dissipation in the first place. Systems of this sort take on a persistent, unique form every time they form, depending on the details of the initial conditions.

as it is a change in the structure of human relationships to one another and their environment. Even so, whether the New Deal is seen as change in behavior or structure is a consequence of how one observes the system, and it is possible to describe even agricultural origins in terms of the same system structure merely exhibiting different behavior. Any set of changes can be seen either as a new system or as the old system exhibit-

Homeostasis—Dynamic Equilibrium

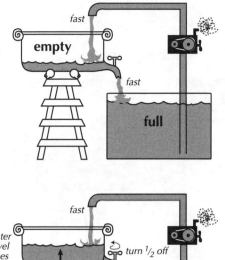

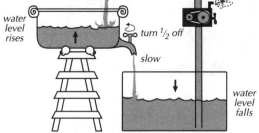

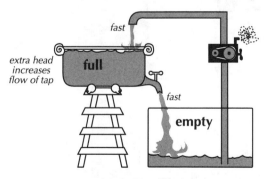

New Equilibrium

Far From Equilibrium—
Dissapative Structure

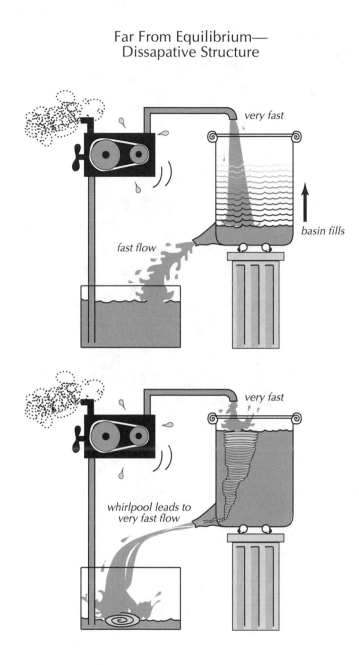

ing new behavior. A particularly clear example of this is the change in social structure that occurred in the Great Depression of the 1930s.

One could look at the New Deal as a change in the structure of American society. Certainly the monied and privileged saw it that way, as did the recipients of the new government handout, Social Security. Many relationships in society changed fundamentally. The United States of America was a new nation, little resembling the one at the turn of the century, where there was no social safety net. Reasonable as this analysis might be, it applies only inside a rather local purview.

With a change in the level of analysis, an entirely different pattern pertains. A unique contribution of the United States at the time of its independence was the sanctity of the banking system. With rare insight the Founding Fathers set up a system that paid all debts incurred by the Colony. The 1930s were times of social turmoil when some countries were turning to fascism while others, like the United States, looked as if they might move catastrophically to the left. In this light the New Deal appears as perfectly normal behavior of the American banking system, as it took appropriate steps to remain stable. The Social Security system was a device engineered to buy off the underprivileged classes and avoid revolution. There was no change in structure here, just a new pattern of behavior in the same old banking oligarchy. Thus whether the change is seen as one of structure or one of behavior is entirely a matter of how the observer looks at the system. Likewise, whether or not a system exhibits stability or instability is not a matter of nature, but comes from how the observer defines the situation (figure 8.6).

Subjective Rules and Objective Laws

The distinction between structure and behavior is important. Howard Pattee has identified that we need two sets of restrictions to describe phenomena. Once again, here the issue is description of nature and not nature itself. The two sets of restrictions Pattee labels "laws" and "rules." *Laws* capture the *dynamical* aspects of phenomena and are the contribution of the observed to phenomena. Laws are rate-dependent, inexorable, universal, and structure-independent. They restrict phenomena to what is possible. *Rules*, on the other hand, are more local and capture the *structural* aspects of experience. Rules are rate-independent, local, arbitrary, and structure-dependent. They capture the linguistic, symbolic aspects of system description. They are the contribution of the observer.

FIGURE 8.6 The monster who is chasing is perceived as bigger and with a menacing face. The pursued monster is frightened and smaller, even on paper. Actually, they are the same image. The context makes all the difference. Illusions of this type were developed by Roger Shepard in *Mind Sights* (New York: Freeman, 1990). This one was drawn by Kandis Elliot. We could not get the illusion to work until the figures were chunky and the space around the "larger" figure was crowded.

Recognizing something as a table relates to rules, not laws. While a table may tip over at a rate, it is not a table at a rate. A symbol relates statically and arbitrarily to that which it represents. The definition of "table" is an arbitrary distinction, made by the observer. Rules relate to the observer, and decisions to observe in a certain manner. They restrict phenomena to what the observation protocol allows one to see. Laws, on the other hand, relate to phenomena by virtue of what is possible. If nature never does such a thing, because it cannot, then you cannot observe it no matter what decisions are involved in the observation protocol. Pigs do not fly, because animals of that size cannot support themselves in air, given the way mammal physiology works. No amount of looking at the sky with any sort of fancy instrument will reveal pigs flying under their own steam, because physical laws deny the possibility.

In the previous section on system collapse we mentioned the period of collapse when the positive feedbacks in the system run wild. At that time there is a suspension of structural rules within the system, and what one sees is an expression of naked law-prescribed dynamics. Only when the system of positive feedbacks encounters new constraints does a new set of rules emerge. Structure reasserts itself, albeit with a new structure. The suspension of the old rules is instantaneous. This is because rules are structure-dependent, and the positive feedbacks that tear down the prior system are a consequence of the very structure that they are dismantling. The rules suddenly have no structure to which they might apply. The positive feedbacks deny the very system that the rules prescribe.

From the perspective of the old system, the new system is a mystery. Even in the process of real time collapse, primitive hunter-gatherers would lack the perspective to see what was happening, as we interpret it with hindsight. They may have found the changes in their society stressful, but would have no reason to suppose that it was all a precursor to agriculture, a system of which they had no knowledge. Cultivation and sedentary living would have seemed most outlandish behavior. If anyone had suggested to them that their immediate descendants were about to do exactly that, first, they would have great difficulty understanding the concepts involved, and second, they would hardly believe the prediction. Gentile Victorians would have been appalled to know how even the most mannerly of late twentieth-century peoples carry on. There is a fundamental difference between considering a process as it goes forward, on the one hand, and, on the other hand, interpreting the unfolding of events once one knows the outcome. This is a constant dilemma for students of any type of historical system. The motives of actors at the time the history was unfolding should not be seen as some effort to get to where we are now, although the temptation to view it in those terms is always superficially appealing.

Contradiction and Dual Description

There is a certain contradiction to description by rules as opposed to laws. The solution to the contradiction is to recognize that both are necessary for a full account of phenomena. These sorts of contradictions are already a part of physics, where photons are both waves and particles. The wave is the dynamical, law-based part of the phenomenon, while the particle is a structural, rule-based consideration. Insisting on a unified description leads to contradiction. If rules pertain to the observer, then a complete account of the phenomenon in terms of rules does not allow the observed end to make its contribution. The observation is of nothing. Conversely, if only laws or dynamics of the observed are invoked, then the phenomenon denies the observer. There is no explanation, because no one looks, no conceptual categories are mapped onto external dynamics, and so no phenomenon is recorded.

The wave/particle duality from physics shows exactly this sort of apparent contradiction. Electrons allowed one at a time through a pair of slots leave marks on the screen behind the slots (figure 8.7). One can see the hits emerge, one at a time, as dots appear corresponding to individual particles. However, as more hits occur, a pattern becomes clear. Surprisingly, it is a wave interference pattern, with most hits occurring between the slots to give the wave pattern its strongest amplitude in that region. This means that each particle, as it made its dot contributing to the pattern, either went through both slots (but how?) or somehow "knew" about the other slot as it went through the one it did (figure 8.8). Neither explanation is satisfactory. The better resolution is to acknowledge that both a wave and a particle description are necessary. Contradiction is avoided because, although dynamical waves cannot be structural particles and vice versa, the two descriptions are kept disjunct. They are used in parallel, not together, and so do not come together to create a contradiction. It is only if the two modes of description are pressed together in a unified account that contradiction arises. Dual

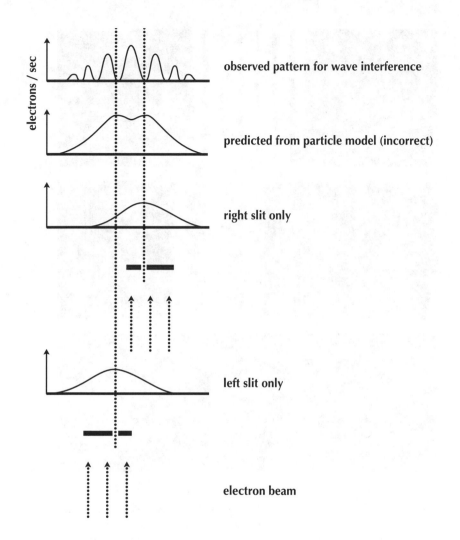

FIGURE 8.7 Fire electrons at a screen through a slot, and one gets an accumulation of hits opposite the slot. Do the same with another single slot, and the same happens opposite that slot. The prediction is that two slots open at the same time would let through electrons in a pattern that is the sum of both patterns individually. That is not what happens. Actually a wave interference pattern emerges, as if the isolated electrons, moving through one slot, know that the other slot is there.

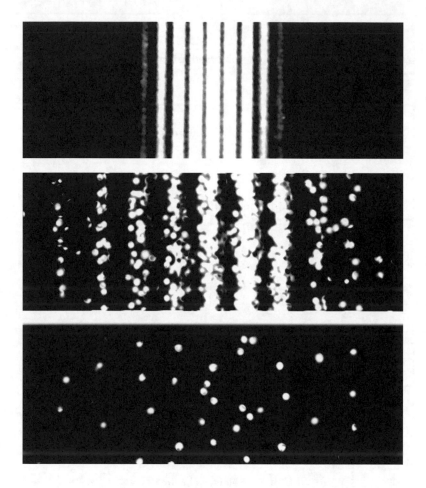

FIGURE 8.8 The electrons show their individuality by making a dot, one
for each electron, on the screen. In the beginning there is no pattern
apparent, but an accumulation of electrons show the wave interference
pattern, as if the particles went through both slots.

description seems the best way out of the dilemma, given the Western intellectual tradition that avoids contradiction.

In biology and the social sciences dualistic models are not favored. Biologists prefer to patch over contradictions as best they can with ambiguous terminology. Biologists usually prefer to erect unified models, with the contradictions swept under the rug. However, a more careful look exposes all sorts of contradictions. The New Deal example is one such case. The question, "Was the New Deal really a change in structure, or was it just a change in behavior?" is not a fair question. Clearly it was both, and we need both descriptions to achieve an adequate account of the phenomenon. There is in all systems descriptions a tension between structural and dynamical description, and it is well to acknowledge that fact. In biology and the social sciences that deal with multilevel systems, the distinction between structure and dynamics is crucial and cannot be avoided.

In both biology and the social sciences there is a historical component, and that is where confusion can arise if one insists on a unified account. Biological evolution and human history both entail the recognition of something that has significance and needs explanation. The distinction here is between different types of explanation. On the one hand there is the question of how a particular situation arose. Notice that this question pertains only after the fact, once the observer knew that it has indeed occurred. On the other hand, there is a separate need to find general principles upon which antecedent dynamics move forward, without regard to the particular outcome that actually occurred.

In evolution, the distinction is between purpose and mere existence. A wing has clear adaptive advantage; it is for flying. On the other side of purpose is the blind evolutionary mechanism of natural selection that knows not where it is going. There is nothing necessary about the emergence of birds as a class of vertebrates. To take another example, in human history the duality is between registering an event, and explaining how a particular historical outcome came to pass. The outbreak of World War I has been attributed to the assassination of an Austrian archduke, but that is not an adequate explanation. The workings of general background processes that happened to lead to this event is also an explanation. Given imperial Britain and Germany in the nineteenth century, war was a likely eventuality. If not the duke, then it would have been something else.

In evolution and history the critical tension is between explaining something once you know it happened, as opposed to understanding the

188   Evolution and Revolution in Creating Complex Structures

general principles upon which the underlying system moved forward, independent of the particulars of the end product. The end product corresponds to empirical structure, and the mechanism that moves the system forward corresponds to structurally independent dynamics. The difference is between looking back from the end product (structural rules), versus describing how the system moved forward and happened to produce any products at all (dynamical laws and blind mechanics). An explanation of a particular situation has hidden inside it the subjective decision that the situation was distinctive enough to be worthy of explanation in the first place. For example, the study of bird flight evolution takes for granted that locomotion through the air is distinctive enough to invite a special explanation. Similarly, historians do not pick a span of years capriciously, but rather base their choices on the significance of certain events that are deemed important, such as World War I.

Notice that there is an implied purposiveness in the products of both evolution and history. Wings are recognized as being for the purpose of flying, and World War I served the purpose of realigning international relations. Evolutionists are well aware that the first steps toward wingness were not driven by the purpose of flying. While there have been arguments that early winglike structures served to stabilize the organism during fast running, other suggestions have been that they were merely butterfly nets for insects scared into the air by a running dinosaur. The "wings" supposedly directed the insects toward the mouth. However, in the final analysis, the achievement of flight appears to be the most significant part of the whole process. From the modern perspective looking backward, flight is the essential purpose of wings.

Similarly in history, the reasons for actions appear different in hindsight from the way they did to even the principal actors. The actions that led to war, like the precursors of the bird wing, were not taken in order to create war. The contemporary record during the first year of World War I indicates that expectations were that it would all be over by the first Christmas. Contemporary accounts of the murder of the archduke that started the whole thing do not indicate an awareness of the cataclysm that was about to happen. Only occasionally is there a telling prediction. The actual consequences of the early stages are set in the context of a large number of other early events that turned out to be insignificant in the end (figure 8.9).

Thus purpose and significance are decipherable only after the fact, and from the perspective of the observer's models and emphases. At the

The Tiger: "*Curious! I seem to hear a child weeping!*"

FIGURE 8.9 A Will Dyson cartoon appeared in the *Daily Herald* as the allied leaders returned from signing the Treaty of Versailles, at the end of World War I. To the year, it predicts World War II, an event not to happen for twenty years. The cartoon has been redrawn in the same style as the original by Kandis Elliot.

time, lawlike dynamical precursors have no purpose or direction; they merely exist without meaning. Therefore there is no need to seek purpose and values in the workings of the material system. Apparent purposiveness comes from the observer recognizing certain structures and events as significant given hindsight. Even so, it is not enough to say that purpose and meaning come just from observer bias, and dismiss it. Without the observer, there is no historical or scientific observation, and the sentient observer cannot work without making decisions about what is important.

Suppress the rule-based observer decisions, and one gets wrong predictions in physics and in the biological and social sciences. Science is not just about the material system. It also necessarily involves scientists deciding, observing, and knowing. In biological evolution contradictions are hidden in weasel words like "adaptation." Adaptation confuses lawlike processes that move a system forward, and observer-based rules that recognize significance and purpose after the fact. Life has emerged with an extremely large set of very particular structures—so particular that a random process, incapable of recognizing purpose, would have too large a search space to have produced what we see. Note the solution to the dilemma is not to assert that the mechanisms are in fact purposive. The solution is to recognize the need for dual description, which distinguishes between after-the-fact purpose and dynamic unfolding of events.

In summary, both science and history require two levels of explanation. There is structure, and it behaves. Structural behavior requires an account of lower-level mechanism and upper-level purpose. Lower-level dynamics drive structural behavior, as when biochemical dynamics cause muscular contraction. That contraction can be seen as contributing to higher-level dynamic, movement, at the level of the whole animal. *How* the muscle contracts can be found at the lower level, but the reason *why* it contracts can be found only in the animal's goals. Was the movement to escape a predator, or to find food, or for no obvious reason at all? When dealing with more than one level of organization, one is forced to use dualistic, complementary, and apparently contradictory models. One model accounts for what blindly happens, and the other is interpretable only in terms of function as recognized by the one who perceives. By dealing with the world in hierarchical terms, and including the role of the observer as well as that of the observed, we can handle a suite of dilemmas that have plagued science across the board.

Conclusion

In this chapter we have looked at how complex systems emerge, and why we might expect them to appear hierarchical. As with other issues raised in this book, we have couched the study of emerging complex systems in terms of the observer and epistemology. The emergence of complex hierarchical systems involves a play between rule-like structure and law-like dynamics. Out of this tension there emerges a requirement for dual description in science, known as the principle of complementarity in physics. The generality of a hierarchical worldview is indicated by the application of principles from quantum mechanics across the spectrum of science to include the social sciences. Explanation includes the observer's conceptual structure, the lawlike dynamics like natural selection that generate structure, and the teleological role that structures serve in the observer's models, developed after the fact of observation.

9

Conclusion

Limits on Prediction

In this book we have presented an epistemology of science that revolves around limits on an observer's ability to know the world in itself. We have also characterized science as the business of making predictions into the future. Given that there are limits on human knowledge, what are the limits on prediction? If we know the behavior of the parts of a system, can we always predict the behavior of the whole? The answer is that, depending on the form of the question asked, some systems resist prediction, even given a large knowledge of the parts. In such systems, the problem is that the whole does not completely constrain the parts, and so there are degrees of freedom for unexpected outcomes. When a system is intrinsically unpredictable, it is because the question implies an unstable set of constraints. When that happens, any one of a large number of parts can take over control of the outcome.

Take, for example, the question, "When I pour champagne into this glass, where will the streams of bubbles arise?" The problem is they will come up in new combinations of places every time. There are many imperfections on the inside of a glass, and the local turbulence of the sparkling wine will produce streams of bubbles on any one of them by accident. The problem is more complicated than just the sum of the imperfections inside the glass, because when a stream of bubbles arises at one of them, it suppresses potential streams at others. It is in the nature of the question that each accident can take over the system at any time. The imperfections cannot be averaged.

On the other hand, this does not mean that one cannot make predictions about champagne in general. It is possible to predict that a good bottle will fizz when poured. This more general prediction can be made because, with molar quantities of carbon dioxide particles (billions and billions of them, more than Avogadro's number), the molecules can be averaged, and the average particle will reliably make its contribution to the fizz.

It is well known that many fine-grain predictive models are inept when used to predict over extended time periods. For example, models for weather cannot be used for climate prediction. The reason that predictions are so wrong in the long run is that the system parts to be used in the prediction are unstable over a longer time window. For time periods of appropriately short duration, weather-predicting models do a remarkably good job. Warm fronts move in a consistent direction, and are sufficiently persistent so that they are useful predictors of weather across the continental United States for a full week. However, fronts are long gone before they can play any role in predicting a mild winter. To predict the sort of winter that is in store, one needs to consider the position of the jet stream, a pattern that persists for even several months. At the seasonal level, warm fronts represent the caprice of turbulence, while at the level of the weekend forecast, they move in orderly procession.

If one is unable to make predictions, then a solution is to shorten the length of the forecast until predictive power emerges. In the case of the weather, if one is using high- and low-pressure regions and warm and cold fronts, over a period of months predictions would fail. If one shortens the time of the weather forecast to a week, then predictive power emerges. The failure occurs over the long prediction because the entities used for prediction have dissipated before predicted weather can come to pass. This has implications for what one might call demonstrably false models as opposed to those that are valid but used for a forecast that is too long.

A helpful concept here is Herbert Simon's notion of near-decomposability. It is possible to decompose experience of material systems, but only to a degree. Reductionism puts much stock in the decomposition that occurs in its reductions. The essential lie in reductionist decomposition is that, for the upper level to function, there must be exchanges between the entities at the lower level. Therefore the reduction is not down to demonstrably separate low-level entities, as faith in reductionism would suggest, but is down to less than fully separate parts. The world cannot be fully decomposed to lower-level entities, for it is only nearly

And the King commanded that the hat he had bought, and all the other hats, too, be kept forever in a great crystal case by the side of his throne.

But neither Bartholomew Cubbins, nor King Derwin himself, nor anyone else in the Kingdom of Didd could ever explain how the strange thing had happened. They only could say it just "happened to happen" and was not very likely to happen again.

FIGURE 9.1 In *The Five Hundred Hats of Bartholomew Cubbins*, Dr. Seuss offers a perfect explanation, but it has no predictive power.

decomposable; otherwise upper-level entities could have no coherence. All this should not be surprising, given the earlier chapters on filters, surfaces, and strong versus weak connections.

It is the near-decomposability of the atmosphere that causes the weather-predicting entities to fail over months but not days. High- and low-pressure systems are not separate, and the former fill up the latter over time. Pressure systems, understandably enough, leak. A perfectly decomposable atmosphere would lead to longer predictions, but it would be impossible.

When a model uses inappropriate lower-level principles, the essential problem is one of rapid dissipation of the lower-level entities relative to the prediction at hand. For example, trees grouped by size in a forest do not correspond to trees grouped by species. However, there is often some correspondence, in that small understory trees will often be of species whose ecology causes them to arrive late to a site, whereas the tall canopy trees in an immature forest belong to different species whose ecology caused them to arrive and thrive early in stand development. Over time, the coherence of trees of one species being the same size class dissipates, because individuals of the same species grow at different rates because of local accidents. The separate classes of trees by both size and species leak across size differences, as some trees of the understory species grow to join the other species in the overstory. This is, of course, the same problem as using an appropriate model for too long a prediction. High- and low-pressure regions leak into each other, as do species-size classes of trees. Thus predicting species composition of a forest by tree size, an inappropriate model, can work for very short periods in young forests. An inappropriate model would be called in vernacular a "wrong" model. A wrong model in this sense is one in which connections between parts of system components are stronger between individual system components than within system components. In this light, there is little difference between a right model used for too long a prediction and a model that is just plain wrong in a vernacular sense.

Even when an explanation involves principles that are unstable for prediction over a given period of time, it may offer prediction for a pared-down forecast. However, if the time frame must be made impracticably short, then one is faced with a valid explanation, but it has no predictive power. This is the explanation offered for the never-ending sequence of hats that appeared on the head of the hero in Dr. Seuss's *The Five Hundred Hats of Bartholomew Cubbins.* "But neither Bartholomew Cubbins, nor King Derwin himself, nor anyone else in the Kingdom of Didd could ever explain how the strange thing had happened. They could only say that it just 'happened to happen' and was not very likely to happen again" (figure 9.1).

In systems whose constraints are unstable, the only solution for restoring predictability is to change the question. Some dynamics are inherently unstable, and no effort on the scientist's part can overcome these limits from the observed. Not only can we not know everything, but we must also accept that not all the questions we ask have solutions that generalize into the future.

Summary

The point of departure for this book has been an attack on the view that science is an objective process concerned with discovering how the material world works. The alternative conception we propose is a dualistic vision of science. We propose that the contribution of the material world is blind, lawlike behavior that generate external dynamics, unconcerned with purpose, function, or description. For the process of scientific testing each day, the world is just out there, doing what it does. It is active, dynamical, and changing. That is the contribution of the observed. Equally important in science is the contribution of the observer. The observer is responsible for recognizing boundaries around entities, erecting definitional criteria for entities, recognizing phenomena, and building models. Contrary to the view of science as a passive enterprise, done by priests in white lab coats, we propose that the role of the scientist is actively to construct models that link observations of dynamics that arise independent of the observer's decisions to meanings internal to the observer. Modifying an ever-expanding web of belief is just as much a part of science as collecting data, and no data can be collected without an active observer making decisions as to 1) how the world to be is measured, and 2) which measurements are considered noteworthy.

There is a story that illustrates the point of there being no such thing as an observer-free observation. It has currency in systems analysis circles, but we do not have a firm reference for it. An American fighter squadron was suffering inexplicably high losses in World War II, and a systems analyst was brought in to find a remedy. After some weeks of collecting data on who repaired and who flew which planes how often, as well as a host of other measurements about damage, the analyst had a solution. The recommendation was to put more armor in certain places. The squadron leader was incensed and said, "I have experience here, and I know for a fact that planes never get hit there." The analyst replied, "No, planes that get hit there never come back." For us, hierarchy theory turns on a self-consciousness about being mortal data collectors, not omniscient beings. Forget for a single moment that data are only codified experience, and one invites the error of the squadron leader. The experiencer must be part of the measurement.

Our account of hierarchy theory flies in the face of naive realism. It is a constructivist model that accepts the existence of external reality, but denies that human knowledge is a direct link between reality and conception. Humans know their concepts and ideas. We do not know the world, itself, directly. As science evolves, it is scientific models that change, not dynamics of the external world. Models do not correspond with external reality in other than an arbitrary fashion, they represent it. Description is different from approximation. Description uses symbols, and all symbols have an arbitrary relationship to that which they represent. By contrast, approximation is always in the same terms as that which it approximates. If we grant that there is a true rate at which a particle moves, approximations are always in terms of velocity, with more and more decimal places. As science progresses, models become more useful and parsimonious, given an interaction between the observer and the observed. Models progressively employ more robust and reliable structures, and come to predict the experience of external dynamics more accurately. That is not the same as becoming more true. Any model, at any time, could be subject to a paradigm shift in which the foundational questions in a discipline change. Paradigm shifts occur not because the true answers to all of the old questions have been revealed but because a working group of observers have changed their questions and emphases. New questions, not new truths, guide the path of science.

In this book, we started by offering an epistemology of science and then went on to discuss 1) identifying levels that may be ordered hierarchically, and 2) a variety of ways for ordering levels into hierarchies. The beauty of hierarchies is that they allow us to explain a wide range of diverse phenomena within one unifying scheme. The phenomena within the hierarchy are stable subunits, which run the gamut from concrete to abstract relations. Hierarchies have integrity because they consist of stable, cohesive subunits that are linked and ordered according to an organizing criterion, which itself may be as concrete as containment, in the case of nested systems, or as abstract as cognitive models in a child's or scientific discipline's cognitive development. Causal relations between levels are not push-and-pull, Newtonian mechanics. Instead, levels are related by what is possible from below, and what is allowed from above (figure 9.2). Context, constraint, filters, response rates, characteristic frequency, and size all link levels, and generate surfaces that isolate entities within levels.

As science proceeds into the twenty-first century, replacing realism and Newtonian causality with a separation between the observer's rules and the observed's laws should help resolve paradoxes that block progress in particular disciplines. For example, John Searle argues that

in order for psychology to progress beyond models that analogically compare humans to computers, a revolution is needed. Just like in the field of biology, where Darwin's theory of natural selection came to replace anthropomorphic and teleological explanation, psychologists need to replace "as if" computer metaphors. Psychology needs to move beyond explaining psychological functions in terms of internal homunculi that perform meaningful computations. The solution involves a differentiation between blind mechanism, which facilitates configurations that suggest structure, and the observer's evaluation of the function that a structure serves. Without that distinction, even the most high-sounding theories translate to little men inside doing what needs to be done.

Function is always recognized after the fact and rests on an observer recognizing the significance of a phenomenon, within a cognitive model. Function is not discovered in nature. It is not appropriate to deal with the issue of function by creating words like "adaptation" that pretend that function is somehow inherent in structure, but rather to separate consciously structure and mechanism from function. It is time to acknowledge properly the observer's contribution to science. This book is offered as a step in that scientific revolution, which we hope the twenty-first century holds. Brute mechanism and naive realism are tired. In order to avoid paradox, the next generation of science needs to acknowledge the scientist's active, constructivist role. The revolution will be an involuted paradigm shift. Instead of the topics changing, it will be the form of the questions that change, even though they will continue to pertain to the same topics (e.g., cancer, mental health).

In this book we have attempted to lay out the basic principles of hierarchy theory and to apply them to enough commonplace examples and across enough disciplines so that the generality of the approach should be clear. Although the questions addressed by the approach may appear sometimes esoteric, there is a great deal of common sense underlying the suggested protocol. By now it should be clear as to the cost of acting on a belief that science approaches some kind of final truth. Our approach

FIGURE 9.2 The great leaves of *Victoria amazonica* could only occur in a water plant, because a leaf that size could not be supported in air. Therefore, the habitat of the leaf is limited by physical requirements coming from below. The physical limits are as to what is possible. From above come the biological constraints as to what is biologically allowed. Gigantic as it may be, it is still constrained to be a leaf. *Photograph by Paula Lerner*

has rested on a utilitarian view of science, in the hope that unprovable ontological assertions can be kept out of the process of doing science day to day. Hierarchy theory is a theory of observation in the face of complex systems. It pretends to be nothing more, but to us who use it as a guide daily, it seems to help keep our feet on the ground; from that firm footing we find we can leap higher as we grasp at the critical questions. The closest epistemology to the one we use is that of Ernst Cassirer. He is not focused on hierarchies per se, but most of what we say appears in accordance with his views. We might have dedicated this book to him, but our shared families took clear precedence.

References

Allen, T. F. H., and T. Hoekstra. *Toward a Unified Ecology*. New York: Columbia University Press, 1992.

—, and T. Starr. *Hierarchy: Perspectives for Ecological Complexity*. Chicago: University of Chicago Press, 1982.

Cassirer, E. *Substance and Function*. New York: Dover Publications, 1923.

Darwin, C. *The Origin of Species* (1859). New York: Washington Square Press, 1963.

Gardner, H., ed. *The New Oxford Book of Verse, 1250–1950*. Oxford: Oxford University Press, 1972.

Helmholtz, H. von. *Helmholtz's Treatise on Physiological Optics* (1867). New York: Dover, 1962.

———. "On the Theory of Compound Colors." *Philosophical Magazine* 4 (1852): 519–34.

Hering, E. *Outlines of a Theory of Light Sense* (1878). Cambridge, Mass.: Harvard University Press, 1964.

Koestler, A. *The Ghost in the Machine*. New York: Macmillan, 1967.

Kuhn, T. S. *The Structure of Scientific Revolutions*. Chicago: University of Chicago Press, 1962.

Pattee, H. H. "The Complementarity Principle in Biological and Social Structures." *Journal of Social and Biological Structures* 1 (1978): 191–200.

———. "The Evolution of Self-Simplifying Systems." In *The Relevance of General Systems Theory*, ed. E. Laszlo, pp. 31–42. New York: Braziller, 1972.

———. *Hierarchy Theory: The Challenge of Complex Systems*. New York: Braziller, 1973.

Piaget, J. *Genetic Epistemology*. New York: Norton, 1971.

————. *Psychology and Epistemology: Towards a Theory of Science*. New York: Penguin Books, 1972.

————. *Psychology of Intelligence*. Paterson, N.J.: Littlefield and Adams, 1963.

————. *Structuralism*. New York: Harper and Row, 1970.

Quine, W. *Word and Object*. Cambridge, Mass.: M.I.T. Press, 1960.

Rosenkrantz, R. D. *E. T. Jaynes Papers of Probability, Statistics, and Statistical Physics*. New York: D. Reidel, 1983.

Searle, J. *The Rediscovery of the Mind*. Cambridge, Mass.: M.I.T. Press, 1992.

Seuss, Dr. *The Five Hundred Hats of Bartholomew Cubbins*. New York: Scholastic Book Services, 1938.

Simon, H. A. "The Architecture of Complexity." *Proceedings of the American Philosophical Society* 106 (1962): 467–82.

————. "The Organization of Complex Systems." In *Hierarchy Theory: The Challenge of Complex Systems*, ed. H. H. Pattee, pp. 1–238. New York: Braziller, 1973.

————. "What Computers Mean for Man and Society." *Science* 195 (1977): 1186–91.

Thurber, J. *My Life and Hard Times*. New York: Bantam Books, 1933.

Vygotsky, L. S. *Thought and Language*. Cambridge, Mass.: M.I.T. Press, 1962.

Whitehead, A. N., and B. Russell. *Principia Mathematica*. Cambridge, Mass.: Cambridge University Press, 1910.

Index